"A TIME TO KILL"

Reflections on War

★ ★ ★

"A Time to Kill"

Reflections on War

★ ★ ★

EDITED BY: DENNY ROY
GRANT P. SKABELUND
AND RAY C. HILLAM

SIGNATURE BOOKS   SALT LAKE CITY   1990

To our students

© 1990 by Signature Books, Inc. All rights reserved
Signature Books is a registered trademark of Signature Books, Inc.
Printed in the United States of America
∞ Printed on acid free paper

94 93 92 91 90   4 3 2 1

Cover Design: Julie Easton

Library of Congress Cataloging-in-Publication Data

A Time to kill : reflections on war /
  edited by Denny Roy, Grant Paul Skabelund, and Ray C. Hillam.
    p.   cm.
    ISBN 0-941214-97-4
      1. Mormons—United States—Biography.   2. Veterans—United States—
Biography.   I. Roy, Denny   II. Skabelund, Grant Paul
III. Hillam, Ray C.
BX8693.T56 1989
355'.008'8283—dc20   89-70005
                                                                    CIP

To every thing there is a season,
and a time to every purpose under the heaven: . . .
A time to kill, and a time to heal; . . .
A time of war, and a time of peace.
—*Ecclesiastes* 3:1, 3, 8

CONTENTS

Acknowledgments ix

Introduction xi

One: Going to War. 1

Two: Combat on the Ground 19

Three: Combat in the Air 67

Four: Killing and Being Killed 115

Five: Over There 137

Six: Captivity 177

Seven: Leadership 219

Index 239

ACKNOWLEDGMENTS

The David M. Kennedy Center for International Studies at Brigham Young University provided facilities and other services during the writing and compiling of this book. Without this support, the project would have been much more difficult to complete.

We express our thanks to the many students who, under the direction of Ray C. Hillam, conducted the initial and some follow-up interviews. Kristine F. Abbott, Kathryn Holm Clark, Stephanie Condie, Linda Felt, Elizabeth Ann H. Lingren, Melanie L. Jenkins, Heather Keele, Frances F. Nickerson, Suzanne Willmore, and Connie Wright performed the essential task of typing much of the text. Scott Burnett coordinated the initial set of interviews. The Charles Redd Center for Western Studies at Brigham Young University transcribed some of the interviews and provided instruction in interviewing skills. The advice and standards of Howard A. Christy of Brigham Young University Scholarly Publications motivated us to improve the quality of the manuscript.

Finally, we owe everything to the veterans themselves for sharing their personal experiences openly and frankly. Their cooperation was based on a desire that their experiences might benefit others. We hope their desire will be fulfilled.

INTRODUCTION

This is a book of memories about war. Although it describes both good and bad, overall it portrays war as an arena of horror and tragedy. As one veteran noted, "No one wins a war. Even the victors lose a part of their humanity through the experience of war." Although all such conflicts involve peculiar sets of circumstances, their worst aspects remain constant. Of the many questions the following reflections raise about war, perhaps the most wrenching is: Is there ever a time to kill?

The oral histories excerpted in what follows are part of an ongoing project on war supervised by Ray C. Hillam. Some of Hillman's students conducted and transcribed more than one hundred interviews. Those interviewed included combat veterans of the first and second world wars, Korea, and Vietnam.

Rather than select a workable number of complete oral histories in their entirety, we drew from sixty of the transcriptions and arranged them thematically. This format offered two advantages: it allowed the most interesting parts of each reminiscence to be included, and it highlighted both the similarities and the differences in the experiences of the participants. We believe that the simplicity of the approach belies the impact of the result.

The first chapter, "Going to War," explores the feelings of soldiers leaving home for military service. Many went willingly, prodded by a sense of patriotism, duty, or love of adventure. Others went only under compulsion. This chapter points to the concern of these men for their families and the often painful adjustment to the rigors and loneliness of military life.

As combat is the crux of war, so the chapters on "Combat on the Ground" and "Combat in the Air" are the core of *"A Time to Kill."* These chapters include fascinating and often shocking episodes experienced by men who participated in some of the bloodiest battles of the twentieth century. The accounts they give describe fighting in frank detail, from machine gun battles between bombers and fighter planes in the cold skies over Europe to mass-wave attacks by Chinese troops in Korea to sudden ambushes in steamy Vietnam jungles. These vivid descriptions help the reader feel the emotions experienced by men in combat.

"Killing and Being Killed" attests to the terrible dilemma soldiers face on the battlefield. "Over There" explores aspects of war outside of combat and includes observations about life as servicemen abroad. As the accounts show, the environment of the battle zone exacerbated the misery they endured. They also point out the tragic effects of war on local people whose homeland is turned into a wasteland.

Among those interviewed are former prisoners of war. Their experiences are collected in "Captivity," which includes accounts from German, Japanese, American, and North Vietnamese POW camps. Featured are veterans of the Bataan "Death March," Germany's Stalag 17, and the Hanoi Hilton.

The final chapter, "Leadership," details the struggle to maintain composure under difficult circumstances. Soldiers who bore responsibility for others describe the awesome weight of knowing their decisions meant life or death for themselves and their men.

We believe *"A Time to Kill"* serves at least three purposes.

First, it preserves the experiences of men who participated in events which have shaped our world.

Second, the accounts describe war not at the diplomatic or strategic levels but through the eyes of actual participants. They provide insight into the nature of war: the terror, brutality, confusion, heroism, adventure, strength of character and sorrow.

Third, *"A Time to Kill"* describes the religious faith of combatants under extreme stress — the majority of whom, in this case, are life-long members of the Church of Jesus Christ of Latter-Day Saints (Mormon) — something lacking from many reminiscences of war. We do not, however, claim to have captured the entire range of religious experience in war.

This book does not explain how or why wars begin; rather it suggests why every effort should be made to avoid them.

We believe that understanding war helps prevent it. Hopefully, the following reflections will contribute to the prevention of future wars.

one
GOING TO WAR

J. Keith Melville,[1] Army Air Forces bomber pilot, World War II

My attitude toward World War II is that it was a necessary evil. I felt this way when I first went in. I wasn't overly "gung ho" to go to war, but I felt a responsibility to do my part. This is why I finally decided to sign up to go into the Army Air Cadet program. Of course, one of the factors was the real possibility of being drafted, even though I was working for a Remington Arms munitions plant and contributing to the war effort in that way.

---

[1] J. Keith Melville was born 22 September 1921 in Bountiful, Utah. After graduating from West High School in Salt Lake City, Utah, Melville studied physical science at the University of Utah and the University of Montana before serving in the U.S. Army Air Forces in World War II. Melville was twenty-three years old when he began his military service. Following World War II, Melville received a doctorate in political science; he is a professor emeritus of political science at Brigham Young University.

C. Grant Ash,[2] **Army Air Forces bombardier, World War II**

We knew what we were fighting for! It may have been stylized, it may have been blown out of perspective, but I don't think any of us questioned why we were fighting. We were fighting to save mankind from the likes of Hitler and Tojo.[3] We were fighting for freedom. We put a value on freedom. We knew that freedom was worth fighting for.

We felt that God was on our side, although it was a little startling to learn while I was a prisoner of war that all of our German guards wore a belt—a big silver belt made of an alloy that looked like silver—with a big swastika on it and the words "*Gott mit uns*," which means, "God with us." Every German soldier felt this just as strong, even to the extent of having it on his belly. That kind of shook you a little. "God's with us, and he's with them, too?"

Lawrence H. Johnson,[4] **Army Air Forces bomber pilot, World War II**

I think there was a fury about the Japanese that was shared by every American. We just had a great feeling of

---

[2] Cecil Grant Ash was born 27 November 1922 in Pleasant Grove, Utah. He graduated from Lehi (Utah) High School and attended classes at Utah State Agricultural College (now Utah State University) and Brigham Young University before his World War II experience. Ash was twenty-one years old and single when he entered combat. After the war he earned a doctorate in microbiology and was chief environmentalist for the U.S. Army Corps of Engineers. He is now retired.

[3] Adolf Hitler (1889-1945) and Hideki Tojo (1884-1948), leaders of Germany and Japan, respectively, during World War II.

[4] Lawrence H. Johnson was born 27 August 1923 in Burley, Idaho. After graduating from South High School in Salt Lake City, Utah, and attending the University of Utah for two years, the twenty-one-year-old Johnson went into World War II. Before retiring as a colonel from the Air Force, Johnson was a professor of aerospace studies at Brigham Young University from 1968 to 1971.

betrayal when we were zapped by them at Pearl Harbor. We also heard what had been done to Allied prisoners. It was the-good-guys-versus-the-bad-guys kind of feeling.

Neil Workman,[5] Marine radio operator, World War II

I guess you'd classify my mother as patriotic. She once said, "I'd rather see all six of my sons go to war and have to give their lives for their country than to have one who refused to go when his country called him."

Peter Bell,[6] Army Special Forces, Vietnam

When I returned from my two-year mission for the LDS church, I was informed that I was going to get drafted into the Army. I didn't want to get drafted. I wanted to choose my own destiny. I knew about the Green Berets, the Special Forces,[7] and I figured that if I had to go to war, I wanted to be the best trained soldier I could be. I chose to

---

[5] Cornelius (Neil) Workman was born 14 July 1924 in Delta, Utah. Following his graduation from Lovell (Wyoming) High School and part of a year at the University of Wyoming where he studied electrical engineering, the single twenty-year-old began his service in the U.S. Marine Corps. Workman continued his education after his World War II experience, and he worked as a school teacher and real estate broker in Salt Lake City. He is now retired.

[6] Peter Bell was born 6 March 1948 in Mt. Pleasant, Utah. After graduating from Spanish Fork (Utah) High School, attending Brigham Young University for one year, and marrying, the twenty-two-year-old Bell went to Vietnam. Bell is a law enforcement officer and a member of the Nineteenth Special Forces Group of the Utah National Guard.

[7] The terms Green Berets and Army Special Forces are synonymous. They are military advisors, mandated to teach counter-insurgency skills, foreign internal defense, unconventional warfare, and direct-action operations such as demolition and sabotage missions.

join the Green Berets and be highly trained to be able to survive a war situation.

Chris Velasquez,[8] Navy combat photographer, Vietnam

While I was living on the farm and working, it dawned on me that there had to be something more to life than this. One day I just happened to be sitting at a local restaurant, which was the local hangout for us in the 1960s, and drinking a Cherry Coke. My friend came in; we began to talk, and I said to him, "There's got to be something more to life than just working, Jack."

He said, "I was thinking the same thing. You know, my dad was in the Navy. Why don't you and I go join the Navy?" That's what started the whole thing.

Robert R. Hughes,[9] Marine infantryman and motor transportation officer, Vietnam

In February 1967, I was only three months away from graduating from Brigham Young University. I was classified 1-A, and I received my draft notice while I was in school. The reason I received my draft notice was that they felt like

---

[8] Cresencio (Chris) Velasquez was born 15 January 1943 in Ohio. After graduating from high school, he joined the U.S. Navy. From 1964 to 1972, Velasquez saw six tours of duty in Vietnam. During that time period he was married and divorced. Velasquez attended nondenominational church services during his first tour of Vietnam; he has since joined the LDS church. Velasquez retired from the Navy and is currently studying history at Brigham Young University.

[9] Robert R. Hughes was born 28 June 1942 in Washington, D.C. After graduating from Spanish Fork (Utah) High School and Brigham Young University, where he earned a bachelor's degree in physical education and dramatic arts, Hughes went to Vietnam as a single twenty-six-year-old. After his combat experience, he earned a master's degree in recreational education and taught LDS seminary. Hughes is currently an insurance and investment salesman in Spanish Fork.

I was avoiding being drafted because I'd switched majors a couple of times and I was in school longer than they thought I should be. So, rather than being drafted, I asked the draft board if they could give me a little bit of time to graduate. They said, "No." So I joined the Marines.

Lynn Packer,[10] Army broadcaster, Vietnam

It was late 1968, and I got caught in a bad situation. There had been a lot of criticism for drafting a lot of blacks and a lot of poorly educated people to go fight the war in Vietnam. Just a few months earlier, they'd changed the drafting rules to put college graduates and those who had reached their twenty-fourth birthday at the top of the list. I got hit both ways. I graduated and had my twenty-fourth birthday the very month the rule went into effect. I was destined to go into the military.

Essentially I was paying for the sins of the drafting system. They should've been drafting college graduates and people who had kids all along. But they hadn't been, and suddenly they decided they'd better start doing it, so they really targeted this particular group. Here I was, recently graduated from college, having launched a potentially great career as a reporter for KSL television in Salt Lake City, married, and a baby on the way. And then I got the draft notice. I can remember I got home late one night from cov-

---

[10] Lynn Kenneth Packer was born 12 June 1944 in Brigham City, Utah. After graduating from Box Elder High School in Brigham City, Packer received a bachelor's degree in broadcast journalism from Utah State University. When his military service began, Packer was twenty-five and married. He and his wife had a daughter who was three months old when he left for Vietnam. Packer has been a television news reporter, a college instructor, a media consultant, and a legal consultant. In addition to consulting, Packer currently teaches communications classes at Brigham Young University.

ering the primary elections for KSL and there was a note from Uncle Sam in my mailbox. It was a great shock.

If I could've found a reasonable way to get out of it, I would've. I gave very serious consideration to going to Canada. I was against the war at that point. It was clear to me that the reasons for the war were largely manufactured by this country's administration. The administration view was that Vietnam was a bulwark against communism and it was keeping communism from spreading in Southeast Asia. But others and I knew it was a waste of time, effort, money, and lives.

Richard P. Beard,[11] Army Airborne, Vietnam

I really felt, and still feel, that communism is enough of a threat to humanity as a whole that it's worth the loss of life to oppose it. Freedom is worth fighting for. I'd served an LDS mission in Chile and had seen the communists at work down there. Therefore, I felt very strongly that the United States was right wherever it got involved somewhere in the world to stem communism.

Edmond S. Parkinson,[12] Army Corps of Engineers, Vietnam

Upon arrival at the training/demarcation site—Fort Lewis, Washington—the Idaho engineer unit I was a mem-

---

[11]Richard Paul Beard was born 24 August 1941 in LaMonte, Missouri. He had graduated from LaMonte High School and attended Central Missouri State College and Brigham Young University, where he earned a bachelor's degree in English, married, and had a daughter before going to Vietnam in 1969. Another child, a son, was born while he was in the army. Beard is now an attorney and a Missouri state legislator.

[12]Edmond S. Parkinson was born 19 November 1939 in Rexburg, Idaho. After graduating from Madison High School in Rexburg, Parkinson studied business at Brigham Young University and Ricks College in Rexburg. Parkinson also married, and he and his wife had three boys (ages eight, six, and six months)

ber of found it was to share its billeting compound [quarters] with an armored cavalry regiment (ACR)[13] from California. This ACR unit's personnel were in the process of establishing a profound refusal to enter the war as an organized unit. There had been numerous complaints from their men since the mobilization announcement had been issued in April. Some of the men had fought their planned deployment with whatever means they could conceive. Many went AWOL and failed to report when the time came. Their worst offense, however, was to sabotage their own equipment. On the road from southern California to Fort Lewis, they put sugar into the fuel tanks of many of their trucks, which ruined the engines. When they finally arrived at the training site, some of the enlisted men would lie down in "peaceful protest" and refuse to respond to the orders given them. I observed a wide disparity between those two groups of men. On the one hand, the Idaho boys were resigned to make the best of the situation, while the California unit seemed compelled to devise means to shirk the responsibility of their oath of military allegiance.

Lawrence H. Johnson, Army Air Forces bomber pilot, World War II

My father and mother were very proud of my being in the Army Air Forces. They were very supportive of America's role in the war and were proud of the idea that their family was contributing to it. I'm sure since I was their only boy they were very apprehensive about what might happen. While I was in the Pacific Theater (I was assigned to the Ninetieth Bomb Group, which had a skull and crossbones on the tails of its B-24s and called itself the "Jolly

---

and a girl (age seven) before going to Vietnam as a twenty-eight-year-old. Currently a computer consultant in Washington, D.C., Parkinson has worked in the military and as a farmer since his tour of duty.

[13] Or vehicular-mounted infantry; a fast-moving unit that is generally transported in armored personnel vehicles.

Roger"), a form letter was sent to all parents by the commander of the Fifth Air Force, General Kenney. His one-page letter congratulated them on what a fine son they had. They really treasured it. It was a real positive thing for them.

E. Leroy Gunnell,[14] Air Force pilot, Vietnam

I knew my wife was a strong and capable woman and could manage in my absence. But even knowing that and having the comfort and assurance from my church leaders, it wasn't easy to leave a single parent with a house full of teenagers and younger ones to care for. So I was concerned about having to leave the family at that time.

I gave each member of my family a blessing and just trusted in the Lord that he'd bless us if we tried to live our lives right. I believed he'd watch over and protect me and bless my wife with the ability and strength to meet situations that would arise.

Kirk T. Waldron,[15] Air Force pilot, Vietnam

I submitted a voluntary statement to go to Vietnam because I felt like I needed to have an experience there and

---

[14] E. Leroy Gunnell was born 22 October 1929 in Soda Springs, Idaho. After graduating from Soda Springs High School, Gunnell received a bachelor's degree in English literature and a master's degree in American literature at Brigham Young University. As a career Air Force pilot, Gunnell went to Vietnam as a forty-year-old father of six (two daughters, ages seventeen and fourteen; four sons, ages sixteen, fourteen, ten, and eight). Gunnell has worked as the administrative assistant for the Brigham Young University Honors Program since retiring as a lieutenant colonel from a twenty-four-year career with the U.S. Air Force.

[15] Kirk T. Waldron was born 24 September 1936 in Tremonton, Utah. After graduating from Bear River High School in Tremonton, Waldron earned a bachelor's degree in business management from Utah State University. Waldron was serving in the Air Force when United States involvement in Vietnam increased in the 1960s. Waldron and his wife had four daughters

I wanted to be a part of it. I felt like I wasn't doing my part to help out, and I was very restless and anxious to do that.

I had four little children, and I had some regret about leaving them and my wife behind. She worked hard with those little children. I had some guilty feelings about leaving them behind, and the whole parenting load to my wife. That's the closest I ever felt to regret. But no, I didn't ever regret volunteering for Vietnam duty.

Ron Fernstedt,[16] Marine infantryman, Vietnam

If there was another war, I'd go in a second. I'm an adrenalin addict. You live faster and harder than you could believe possible. You make the closest interpersonal relationships you can imagine, then watch a man die with "Sorry about that," and drive on.

I do not extol the killing, the technology, and the horrors, but I do enjoy the challenge. It's like playing chess with live pieces, or football for the ultimate stakes.

Howard A. Christy,[17] Marine infantryman, Vietnam

Probably like many career officers at the time, I volunteered for combat duty in Vietnam. After all, the United

---

(ages eight, seven, six, and one) when he began his tour of duty in 1967. Waldron was professor of aerospace studies at Brigham Young University for three years before he retired in 1984. He is currently the deputy director of the Utah Department of Administrative Services.

[16] Ron Fernstedt was born 15 March 1944 in Seattle, Washington. He graduated from Lytton (Iowa) High School. Fernstedt joined the LDS church in 1965, a year before he went to Vietnam. A law enforcement officer, Fernstedt is in charge of facilities security for Utah County.

[17] Howard A. Christy was born 9 May 1933 in Berkeley, California. He graduated from Highline High School in Seattle, Washington. Christy, who joined the LDS church in 1962, earned a bachelor's degree in forestry at the University of Washington in Seattle. He was married and had a seven-month-old daughter

States had been at peace for over ten years and many of us were at mid-career and wanted to test ourselves at what we'd been training for so long. I was a fairly senior captain and itched for the opportunity—of course having no real idea of what I was in for.

I had seen the developing conflict differently than most. Being the East Asia intelligence briefer at Fleet Marine Force Pacific headquarters in Hawaii,[18] I saw the ugly strategy of terrorism on the part of the Viet Cong unfold, grim and tragic episode by episode, and had, along with a lot of others, strong feelings that someone should help the South Vietnamese hang on to what little freedom they had left. When one of my colleagues at FM Pac, a captain who had gone out to Vietnam to be an advisor to the South Vietnamese army, was reported as having been captured, tortured, and dismembered by the Viet Cong, I had no doubt that I wanted to go.

I went. In fact I cut a plush Hawaii tour short by a year to do so. The glow disappeared the first day in the country. Even though I was unable to go any closer to the front lines than the division headquarters (owing to security restrictions that applied to my intelligence billet in Hawaii), within hours it was painfully evident that I was in for a grim experience. On the third day, and my first day as division intelligence briefer, just before I was to set up my map in front of the commanding general, a staff officer hurriedly entered the briefing room and laid a slip of paper in front of the general. General Walt was visibly shaken by what he read. Silently he pushed the report in front of the other

---

when he went to Vietnam at age thirty-two. Retiring as a lieutenant colonel from the Marines, Christy earned master's degrees in library science and history. He is senior editor of Scholarly Publications at Brigham Young University.

[18] The command headquarters of the Marine Corps in the Pacific, under the command of a three-star general.

generals at the table, and the slip then moved quietly up and down the table for all the senior officers present to read.

What the message said was that Lieutenant Colonel Ludwig, a personal friend of General Walt, had, at his newly established command post at Hill 55 a few miles to the southwest, tripped and been killed by an ingeniously contrived booby trap—a trip-wire detonator attached to a live 155-millimeter field-gun shell buried in the ground. The deadly device had exploded with a roar. An eyewitness reported that after the explosion he saw the upper half of Lieutenant Colonel Ludwig, his lower extremities blown away, gruesomely sliding into the large crater the exploding artillery shell had created.

This tragic episode corroborated the comments of another senior Marine officer quoted in *Time* magazine a few weeks earlier. He was the commander of a helicopter detachment assigned to aid the South Vietnamese army. While on a flight he'd been hit by ground fire. He lost both *his* legs too but was still alive. He simply said that Vietnam was a "humorless war."

But I still wanted to have my chance to command a unit in close combat. A few weeks before my opportunity came, a major, who had just been reassigned from a battalion in which a platoon had been caught in a massive ambush and annihilated, was assigned as my immediate senior in the division intelligence staff. When I told him I intended to command a rifle company as soon as I could, he looked up disdainfully and said, "You're a damn fool."

Ivan A. Farnworth,[19] Army infantryman, World War I

I was inducted into the service at Camp Fremont, California. I spent three months there. When we got to

---

[19] Ivan A. Farnworth was born 19 January 1897 in Chester, Utah. After graduating from Blackfoot (Idaho) High School and attending Ricks Academy (now Ricks College) in Rexburg, Idaho,

Camp Miles, they put us in a big shed. I had the flu. At that time, the flu of 1918 was killing a lot of people. In fact it killed more than the war did. A doctor put a thermometer under my tongue. I was burning up with a fever, but I just rolled the thermometer up on top. I didn't want to be sent back. Our captain was turned back on account of the flu and he cried like a little child. He felt so bad that he couldn't go over with the troops. So I just rolled that thermometer up on top of my tongue. The doctor came and said, "This isn't working, put it under your tongue again." He went off to pick up another thermometer and I rolled it back up. He said, "You are sick, aren't you."

I said, "No, I feel fine."

He said, "How come your face is so red?"

I said, "It's always red. I just have a naturally red face."

My squad could see how bad I wanted to go. Those kids just covered me up with their overcoats and carried my stuff onto the boat.

Ted L. Weaver,[20] Army Air Forces bomber pilot, World War II

The first day I was there I went to the induction sergeant in my barracks and told him I wanted to see the Air Corps liaison officer with intentions of becoming one of the Air Corps personnel. He said, "Okay, you can see him.

---

a single twenty-one-year-old Farnworth began his military service with the U.S. Army. He worked as a railroad inspector and businessman. Farnworth died in 1989.

[20] Ted L. Weaver was born 16 December 1921 in Idaho Falls, Idaho. He graduated from Idaho Falls High School and studied music and pre-dentistry at Brigham Young University and the University of Utah before beginning his military preparation at age twenty-two. Weaver was engaged when he entered World War II. After his military service, Weaver earned a master's degree in physics and pursued a variety of science and business interests in Salt Lake City. Weaver died in 1986.

I'll put you down." He wrote something on a piece of paper by his desk and indicated for me to leave. So I left his desk and went around and got right back into the line that was lined up in front of his desk and waited my turn to get up to his desk again. I told him my name again. I said, "I'd like to know what time my appointment with the Air Corps liaison is. You said I had an appointment. What time is it and what day?" He acted surprised and angry or disgusted — I don't know which — and rattled off, "One o'clock next Tuesday." I thanked him and got out of line again. Then I went on about my duties. This was toward the end of the week.

The following Tuesday, he called all the troops out and started giving them their details involved with being a new draftee: KP duty, wheelbarrow duty, latrine cleanups, and the things that were assigned to keep the guys busy. He called my name up as one of the cleanup details. I was supposed to go out on cleanup with a wheelbarrow and clean the area up. I spoke up and reminded him that I had an appointment to see the Air Corps liaison officer. "Oh, that's right," he said, and he excused me from the detail. At a quarter to one, I left the compound and went up to where the liaison officer's office was. I sat on a bench in the barracks where he had his office. I waited awhile. There were two gentlemen sitting there when I got there. They went in and they came out. Three or four more came in and left. I waited there until 4:30 p.m. and he still hadn't called my name. A runner-messenger came in and asked for me by name. He said, "Get on back to your barracks area. You are shipping out."

This aggravated me to no end. I just told the fellow, "Well, you'll just have to wait. I have an appointment to see the officer in there."

By then no one was coming or going, and he was just sitting in his office. I just got up and went in and stood at attention in front of his desk. He didn't even look up. After a minute to a minute-and-a-half, he finally put his pencil down and looked up and said, "Well, what do you want?"

I gave him my name and serial number and told him that I was told that I had an appointment with him at one o'clock to see about getting into the Air Corps.

He looked down at his desk and looked at all of his appointments and he said, "I don't have any Weaver on this list."

I said, "Well, I was told that I had an appointment with you, and now I've been informed that I'm shipping out."

He said, "Well, if you are shipping out, it's too late to do anything about it. You'd have to have letters of recommendation and photocopies of your birth certificate and a bunch of tests."

I had an envelope in my pocket that had all the information I'd obtained in preparation for getting into the Civil Air Corps up to and including the appointment for the physical. I reached in my pocket and dropped it on the front of his desk and said, "I have it."

He opened the envelope, pulled the contents out, and started thumbing through them. He said, "Well, well." Finally, without even looking at me, he picked up the telephone and called the barracks where I was staying, told the sergeant who he was, and said, "Scratch Weaver, he's in the Air Corps." So they didn't ship me out. Then I had to take all the tests over again. But I passed them and was subsequently accepted as a cadet. This was how I got into the Air Corps.

Martin B. Hickman,[21] Army infantryman, World War II

When you are nineteen, there's a kind of excitement about going to war — the change in your work, and I was

---

[21] Martin B. Hickman was born 16 May 1925 in Monticello, Utah. Hickman graduated from Logan (Utah) High

escaping. I felt this was a sense of liberation; I had no ominous feelings. I was traveling with quite a few Utah boys who I entered the service with at Fort Douglas in Salt Lake City, so I wasn't alone. And when I got there, one of my high school friends was the company clerk in a training company, and I was able to have, or re-establish, that contact. The interpersonal relationships get established pretty quickly. I never had the sense of being lost until I left basic training and went overseas. I had one friend, but he was in another group. From the time I left basic training until I got to my combat outfit, I was alone — a replacement. I was really disoriented, just sort of in a dream world. The thing that saved me from the disorientation is that I read a great deal; I found solace in books.

It took me about a month to decide that I wasn't cut out to be in the military. I didn't like it; I didn't like the regimentation; and I developed a fairly cynical attitude toward military service, particularly during basic training. While I was in basic training, I was selected as a squad leader. One day the platoon lieutenant said to me, "Hickman, you make 'sir' sound more like 'son of a bitch' than any other man." I guess from then on my expectations were just to get out; getting it over as quickly as I could. I didn't apply for officer's training school; I was just willing to live it out as a GI.

School and studied pre-law for two quarters at Utah State Agricultural College (now Utah State University) before serving in World War II as a single nineteen-year-old. After returning home, Hickman earned a doctorate in political science. He worked as a U.S. Foreign Service officer for seven years. Currently a professor of political science at Brigham Young University, Hickman served as dean of BYU's College of Social Sciences for twelve years and dean of the College of Family, Home, and Social Sciences for five years.

Ted L. Weaver, Army Air Forces bomber pilot, World War II

I remember the first night in the barracks. There were sixty draftees in the barracks. I undressed and knelt down by my bunk to pray. The place was noisy and full of smoke. The guys at the far end were playing poker. As I knelt down, one of the soldiers nearby said, "What the hell are you doing?"

I paused and said, "I'm saying my prayers, if you don't mind."

He shouted at the guys at the other end of the barracks, "Pipe down, you guys. Weaver is saying his prayers." I don't know whether he did that to embarrass me or out of consideration, but that was the first shock. After that I learned to pretty much ignore the language of the fellows and their attitudes toward me.

Danny L. Foote,[22] Marine artillery, Vietnam

Boot camp was a depressing experience—being away from home. It wasn't the first time I'd been away from home, but it was probably the most traumatic time. I came to a point where I was really questioning why I was in the service, why we were in Vietnam, and if I was in fact going over there, what was I going to be doing, and was I really committed? So when I'd pray about it, these are the types of questions I'd ask. One time while I was in the mess hall it seemed like reality just kind of went away; it was like I was the only person in the room. I had a very peaceful feeling about going to Vietnam from that point on. I went through my combat training without hesitation and without reservation; I just knew that no matter what happened, things were going to be all right.

---

[22] Danny Lynn Foote was born 19 June 1950 in Salt Lake City, Utah. At age nineteen, after graduating from Amos Alonzo Stagg (California) High School and attending a quarter at San Joaquin Delta Junior College in Stockton, California, Foote went to Vietnam in 1969. Foote has worked as a carpet installer. He is currently an electronics technician at Signetics in Orem, Utah.

two

COMBAT ON THE GROUND

Danny L. Foote, Marine artillery, Vietnam

Combat is like no other experience you could ever have in your life. The first time I came under fire, it was like everything that the marines had trained me to do, I did — all of a sudden it was like I was a fine-tuned machine, and I worked. And because I worked, I was proud. And yet after it was all over, the anxiety and the fear that you weren't able to manifest at the time, because you couldn't let stuff like that get in your way, became manifested. It made me physically sick to the point where I'd just throw up.

Hyde L. Taylor,[1] Army Airborne, Vietnam

When you are in a combat situation things happen

---

[1] Hyde L. Taylor was born 30 October 1937 in Salt Lake City, Utah. Taylor graduated from Brigham Young High School and attended Brigham Young University and the College of Southern Utah (now Southern Utah State University) before he went to Vietnam. Taylor left his wife and two daughters (ages ten and nine) when he left on his tour of duty as a thirty-year-old. Taylor retired as a sergeant major from the U.S. Army in 1981. He currently works in property management.

so fast. You don't have time to really think about it. You really only have time to react. I think the people who stop and think about it don't come out. You think about it a lot afterwards and think about the mistakes you made. You react to the situations. I never even got scared until it was mostly over.

I think the times I got most frightened were when all of those things were over and it was a calm time, when you could sit down and have some good security and you could relax a little bit. Then you think back over those things. Then you let the fright come into your life a little bit.

I always said that you could look at somebody and tell how long they'd been there just by looking at their eyes. If somebody had just come into the unit or into the country, their eyes were soft and kind of searching. Somebody who had been there for a long time had eyes that were hard, penetrating. You could also see that reflected in their personality.

Pat Watkins,[2] Army Special Forces, Vietnam

Every day was true adventure. It was an adventure

---

[2] Pat Watkins was born 20 October 1938 in Waynedotte, Michigan. He graduated from Sullivan (Indiana) High School and attended the University of Maryland and Oceanside Junior College before joining the Army Special Forces. Watkins was married, and he and his wife had a daughter (born in 1964), before he began his first tour of duty in Vietnam in 1965. A second daughter was born in 1967. Wounded after ten months of his first tour, Watkins returned to Vietnam for a second tour beginning in October 1967. Between his second and third tours in Vietnam, Watkins received a bachelor's degree in political science from the University of Massachusetts at Amherst. A second battle wound shortened Watkins's third tour. Watkins, a life-long member of the Catholic church, retired as a master sergeant from the U.S. Army in 1980. He worked for three years as the athletic equipment manager at the University of Utah before assuming a posi-

in life and death. I liked that mode as a young man. As I grew older, I still liked it. I never regret one day I spent in the service. There were many times of joy and laughter. Even in combat you could find the funny side of things. There were also times of frustration and emptiness. Emptiness is felt everywhere, but more so in military life, because of the life-or-death environment.

I didn't have time to be scared. I was too busy in a leadership position, keeping other people alive. I think people who are in a leadership position don't suffer fear of combat as much as people who are being led. When you are being led, you can worry just about yourself. When you start worrying about yourself, you might do something dumb. When you have to worry about X number of men, you have to worry about maneuvering them, fighting, and doing the right things to perform your mission. You are so busy that you don't have a lot of time to be scared until it's all over with and you are out. You can sit down and be scared about it later.

Wayne A. Warr,[3] Army infantryman, Vietnam

It's difficult to describe the physical and emotional shock of being hit. Of course, the physical shock was great. It was like having someone hit you with a baseball bat with a full swing as hard as they can. It was a powerful jolt. I

---

tion at Hercules Aerospace in Salt Lake City; he is currently a supervisor of documentation and training at Hercules.

[3] Wayne A. Warr was born 17 September 1946 in Payson, Utah. He graduated from Grand County High School in Moab, Utah. Warr began his military experience as an eighteen-year-old. He did a TDY (temporary duty tour) for three months in Vietnam in 1965. After a second tour of duty in 1967, Warr got married and had a son, who was six months old when Warr went on his third tour in 1970. After twenty-two years in the U.S. Army, Warr retired as a sergeant major. Warr is now a letter carrier with the U.S. Postal Service in Provo, Utah.

don't remember the pain, I just remember the jolt. It actually knocked me down. It hit me in the arm and actually knocked me down right on my back, so there was the shock of that, a feeling of helplessness. I remember that one of my squad leaders got to me right after I got shot, looked at my arm and put a dressing on it over my shirt and everything. It probably took fifteen seconds. He said, "You are going to be all right." With that reassurance I was back under control again. But I can remember a feeling of helplessness at that point: "How bad was it?" I really didn't know.

David L. Evans,[4] Army infantryman, World War II

The infantry was a good place to be snobbish because we could look down on everybody: "They don't know what it's like." Then we saw tanks getting hit by fire and burning. The first time I looked into a tank that had been hit by an anti-tank shell, I became glad I was in the infantry, because it drilled a three quarter-inch rod right through the armor plating and ricocheted around inside. Everybody inside was really chopped up terribly. I think that was one of the worst sights I saw during the whole war. We got so we didn't envy them too much. And then one of my friends was on a ship that got sunk in the Pacific and I stopped envying him, so it kept going. We envied the Air Force because of the way they slept at night. We'd watch them go over, twenty-four planes at a time, and then come back, twelve at a time. That wasn't too hot, either. The one thing that irritated me more than anything else was getting mail from

---

[4] David L. Evans was born 13 June 1925 in Billings, Montana. He graduated from Pocatello (Idaho) High School. After spending a semester at the University of Idaho—South Branch (now Idaho State University), where he studied meteorology, the nineteen-year-old Evans joined the United States' World War II effort. A Presbyterian at the time, Evans subsequently joined the LDS church in 1948. He also earned a doctorate in English and currently is a professor of English at Brigham Young University.

people who'd say, "Oh, how I envy you being there in combat. Here I am, stuck back here in the rear echelons. I'd give anything to trade places with you." I was sorry we couldn't trade places.

Martin B. Hickman, Army infantryman, World War II

After I got into combat and became part of a group, I found a very warm camaraderie. There's a system about combat that develops very supportive relationships. After you have shared the danger of death with someone and seen some of your comrades get killed, there's a drawing together and a supportiveness that's exceptional. It's hard to describe the kind of emotional attachment that you get with these people, people you wouldn't have anything in common with in civilian life.

Norman Wade Sammis,[5] Marine helicopter pilot, Cuban Missile Crisis

During the Cuban Missile Crisis, I was a First Lieutenant flying heavy helicopters out of New River, North Carolina. We were aboard the *USS Boxer*, a converted World War II Essex-class carrier.[6] When you operate from a carrier you carry a Smith & Wesson .38-caliber revolver. The

---

[5]Norman Wade Sammis was born 5 October 1937 in Hackensack, New Jersey. After attending Hackensack High School, Sammis earned a bachelor's degree in engineering from the U.S. Naval Academy in Annapolis, Maryland. As a married thirty-three-year-old Marine, Sammis participated in the Cuban Missile Crisis. At that time, Sammis was a member of the Methodist church; he joined the LDS church in 1971. Since his retirement as a major from the Marine Corps, Sammis has worked as the manager of End-User Computing Support for the LDS church.

In October 1962 the United States and the Soviet Union collided over Soviet placement of missiles in Cuba. The missiles, which would have reached the U.S. mainland, were dismantled after an American threat of force.

[6]A medium-sized vessel not built to be an aircraft carrier but converted into one.

reason you carry a revolver is it can be loaded with tracer rounds so if you go down at night in the water, you can use them to signal the rescue helicopter. Well, we got aboard ship and wanted them to issue the ammunition for the revolvers, but nobody could find it. It was loaded aboard another ship by mistake.

So here I am, getting ready to go to war in Cuba. We've been briefed and all that and nobody has got any bullets. I remember thinking, "This is the most screwed up organization I've ever seen." I think that's when you say, "Do they really know what they are doing?"

Timothy Hoyt Bowers-Irons,[7] chaplain, World War II and Korea

Bravery is the cheapest thing in a war. Almost everybody is brave. It isn't something you need to be proud of. You are or you are not. We are all frightened. I never met a man who wasn't scared of being shot at, but we are taught how to manage our fear. We are afraid of being shot. You can get so steamed up that you forget for awhile. You can even get to the point where you are reckless.

We had a little Mormon boy in Korea who got the Silver Star. He should've had a whole handful of them. I looked him up and talked to him about it. He said, "I didn't even know what I was doing." He got up and wiped out three

---

[7] Timothy Hoyt Bowers-Irons was born 2 October 1915 in Nephi, Utah. After graduating from Juab High School in Nephi, Bowers-Irons attended Snow Junior College and Brigham Young University, where he earned a bachelor's degree in psychology. Bowers-Irons and his wife had a six-month old daughter when he was drafted into the army in 1943. Bowers-Irons was commissioned as a chaplain in 1944 in the European Theatre of World War II. He also saw active duty seven years later in Korea. After retiring as a lieutenant colonel from the U.S. Army, Bowers-Irons ranched near Nephi for several years before poor health forced him into complete retirement.

machine guns, single-handed, and knocked off fifty gooks[8] or something. It was really a tremendous feat. He said the last thing he remembered was that he was behind a rock down on the hillside, and he said the next thing he knew he was up there all by himself with dead gooks all around him. I don't think he was wounded, as I recall, and he got the Silver Star for it. He didn't know he was doing it. He just flipped out of his mind. He went berserk. They gave him a Silver Star and we praise him and say he was a brave man. He didn't know whether he was being brave or anything else. He just blacked out and went up and did the job.

David L. Evans, Army infantryman, World War II

We had one fellow in our outfit who was one of the simplest humans you can ever imagine. He never realized there was any genuine danger in combat. I was walking down the street of a village one night, and I heard a mortar shell coming in and I could tell from the sound of it that it was going to be within a hundred feet. So without any kind of thought, I was down in a doorway just squashing as thin as I could, and he was still walking down the street, and the shell went off just where I expected it to. Shrapnel flew around and knocked sparks off all the buildings around him. He turned around and looked back and said, "Gee, that was close, wasn't it?" And when he saw me, he said, "Oh, gee, did you get knocked down there?" And of course he went through the war unscathed.

Everybody else in our outfit got wounded at one time or another. A couple of them got killed. He was untouched. Our sergeant was also untouched because he knew how to delegate authority. The sergeant never left the command post.

---

[8] A derogatory term for an Oriental, also called "dink" or "slope."

Ivan A. Farnworth, Army infantryman, World War I

Two of my cousins came back in caskets. They never did make it. I saw one of them before he left for the war. He said, "I don't know. I'll never be back."

I said, "Don't feel that way."

But he kept on saying "I'll never be back."

He came back in a box. He got it right in the forehead. He knew he wasn't coming back.

Another cousin came back shell-shocked. He didn't know what he was doing. There were lots of them over there in France. The tears would run out of their eyes. They were just in awful condition. You can't imagine it. They didn't know where they were. They'd pick up anything and put it in their mouths. They had to be watched all the time as if they were animals. Their nerves had gone, absolutely gone. There was a tension on your nerves during the war. Your nervous system is just like an electrical system; it will just stand so much and then it breaks.

Albert B. Haines,[9] Army infantryman, World War II

I was sent out on night patrol preliminary to the battalion attack a couple of days hence. I identified two machine guns, one of which fired at us. We could see where the other one was. Machine guns usually come in pairs; rarely do they come in threes. They cross fire and protect each other, and they are usually protected in turn by foot troops and individual soldiers.

We accomplished what we were sent out to do on that particular patrol. It was no surprise to me (although I didn't relish it) to get the assignment the next day to be the

---

[9]Albert E. Haines was born 23 July 1922 in Pocatello, Idaho. He graduated from Carson High School in Carson City, Nevada. When he began his service in World War II at age twenty-one, Haines was married and had a one-year-old son. After retiring as a colonel from the Army, Haines was the director of space utilization at Brigham Young University until he retired in 1987.

attack platoon for that particular area. We made a very, very early start—it was like four in the morning, I guess. We wanted to get as far as we could—as I recall it was either a waning moon or lots of starlight with the high and heavy snow. We'd wrapped ourselves in sheets. They called them snow suits, but they were nothing more than sheets.

I headed forth with my platoon. We knew where the machine guns were. I deployed my platoon firepower as best I could with the bazooka at a certain position and the rifle grenade launchers at other positions. I focused pretty well on the two machine guns and the shots that we had.

We came down a little closer and saw that we'd been detected because we could see a trail and tracks in the fresh snow where somebody had pulled back from a listening post. So we knew this attack was no longer a surprise to anybody.

There was a burst of machine gun fire from a different direction. That burst killed one person, seriously wounded three of us, and superficially wounded four or five others. Then the two machine guns just kept firing at us. It's what you call being "pinned down." My first thought on being hit was, "It wasn't supposed to happen." But it *had* happened, and there I was. There was little that I could do in the condition I was in (a bullet through both thighs), but we did extricate what few we could and got down behind the hedge with the rest of the troops. I wanted the reserve squad to flank with the bazooka and try to neutralize one or more of the positions. The squad leader was a little more respectful of the fire than he was of me, so not very much happened. It was four or five hours later before another company was able to flank the positions and neutralize the machine guns and take us out. That was the end of the war for me.

Pat Watkins, Army Special Forces, Vietnam

We took off from an aircraft carrier in the Tonkin Gulf and went into the Son Tay prisoner of war camp. There was a Chinese engineer unit there when we went in. They were disorganized. They had their weapons locked up in the arms room. We were shooting them like fish in a barrel. The only casualty we took was a guy who shot himself in the foot. We captured seven Chinese. They were later used in an exchange for some pilots who had been shot down over an island that was part of China. The operation wasn't completely unproductive because we did get some people that our government could barter with to get our people out.

Spencer J. Palmer,[10] chaplain, Korea

I was eating at the mess hall one day and over the loud speaker I heard, "Chaplain Palmer, report to the emergency room." I went to the emergency operating room of the Twenty-first Station Hospital. As I reached that room a handsome, fine young man, a dark-headed kid, was dragged off a helicopter that had just arrived. There was blood all over the place. It turned out that he'd been shot accidentally by a friend in a night security run. I remember the young man lying there on that table, grabbing my hand and yelling out, "Chaplain, I don't want to die. I don't want to die. I'm afraid of death. Just think, I won't see my mother again. Save me, Chaplain. I don't want to die." Then he died.

David L. Evans, Army infantryman, World War II

After the Battle of the Bulge, we moved across the Saar River into Germany. We were nearing a crossroads in

---

[10] Spencer J. Palmer was born 4 October 1927 in Eden, Arizona. Following high school, Palmer attended Eastern Arizona College and Brigham Young University, where he earned a bachelor's degree in fine arts. Palmer, who was single, served as a chaplain in the U.S. Army in the Korean War. After his military service, Palmer received a doctorate in history. A professor of religion at Brigham Young University, Palmer is currently on leave; he is serving as the president of the Seoul (Korea) LDS Temple.

a small town when I heard the shell coming straight toward us. I felt it was a 105-millimeter shell that would fall short, and then it kept on whistling and I thought it must be a 120-millimeter mortar that must be going over. And then it dawned on me, when it was too late to move or anything else, it's a 150-millimeter shell and it's coming straight in. I suppose the only thing that saved me was the fact that it must've hit a wire across the road, because it went off about thirty-five feet in front of me and about fifteen feet high in the air. I didn't hear anything; I just saw the big orange splash with all—well, the paintings you see of the infernal smoke mixed in with all that lurid flame—and that's just what it looked like. I could see the smoke, I could see the flash of the flame, and with my peripheral vision I could see sparks flying across the street right in front of me and all around, and then everything just cut off totally at that moment.

Then I realized that I was on my back looking up at the bottom of a trailer. I tried to figure out where I was, and then I thought, "Well, I'm okay. I don't feel anything, so I must not have been hurt." And so I was lying on my back, and I reached down with my hand and put it on my leg and three fingers went down inside, and then I realized you don't always feel it when you get hit. So I felt around and I could touch a piece of shrapnel about the size of a quarter. What I found out later was that the shrapnel that had hit me was larger than a silver dollar, but it was spinning when it hit. It went down and touched the bone and came back out and almost came outside the skin.

I looked around for Cone, my buddy, and the prisoners we'd been guarding. Then I saw that the prisoners had been right under the shell when it went off. Now there was nothing. They looked like just big blobs of hamburger out there in the middle of the street. Cone was lying about ten feet away from me. He'd taken a piece of shrapnel in the ankle. He wasn't hurt too seriously, but he was convinced that he'd been disfigured—this was the thing that he'd talked to me about for weeks before. He'd carried so many maimed soldiers around, and he said he'd rather die than be maimed, and he was convinced that both legs were blown off. He couldn't feel anything and he was so convinced that he went into shock. A couple of guys ran out from a nearby house to help us. They were running a large radio in there and heard the shell go off. They didn't see us at first, but then they spotted us and one of them rushed over and kept telling me, "Help me carry this guy in," and I kept saying, "I can't, I'm wounded."

"Come on, get up and help me carry this guy."

And I kept saying it and saying it, and finally I grabbed his hand and pushed it down into my leg, and he said, "You're hit. Why didn't you tell me?"

So he got somebody else to help him with Cone and they called for a jeep to come up and take him back. They had one other badly wounded fellow about a quarter of a mile away, and they put the two of them on the jeep. They didn't have any room for me, so I had to walk to the aid station. I don't know how I ever did it, but I didn't feel anything at the time. There was just enough shock in the leg to tighten the capillaries, so I didn't bleed. The shrapnel had also severed the main nerve down the front part of the leg so I didn't feel anything.

I was able to walk back, and I suppose I would've gone all the way back to the hospital without any problem if Cone hadn't died. In the ambulance on the way to the hospital I realized the hand I was holding was turning cold. I was sitting there holding his hand, and when I'd realized what had happened I fainted. When I came to I was back in England.

Lincoln R. Whitaker,[11] Army infantryman, World War II

I remember an occasion when we were in a foxhole.

---

[11] Lincoln R. Whitaker was born 12 February 1920 in Willard, Utah. He graduated from Box Elder High School in Brigham

Two of us were guarding together. A man named Betts and I were talking about life and death. He was from Pennsylvania. He had emotional problems from time to time. On this particular occasion they'd sent a German patrol out and the patrol had come very close to us. We were forced to use grenades to drive them off. We didn't dare fire our rifles because they could see the muzzle blast of our guns and zero in on us. We would've been helpless at that point. So we did everything with grenades whenever a patrol would come into our lines.

We drove the patrol off, and as we sat there pondering what might have happened, Betts went crazy. He jumped out of his foxhole without his gun and started running toward the enemy lines. He said, "I'll kill every one of those dirty S.O.B.s." It fell my duty to jump out and try to save the man, because he was running right towards enemy lines approximately two hundred yards from us.

I know I chased him a good fifty to seventy-five yards before I caught him. Of course, we didn't know whether or not there'd be mine fields out there. I gave him the old football tackle and took him down to the ground. I physically took him back to the foxhole.

Later I took him back to company headquarters and told the captain to send him back to the states for mental and physical evaluation. They sent him to a hospital, but the doctors always sent him back to us. I don't know how many times I sent that poor guy back. He wasn't physically and mentally adjusted to accept combat. He was a menace and a danger to all of us who had to serve with him, even though it wasn't his fault. He just couldn't handle it mentally. The army didn't recognize those things. They would

---

City, Utah. When he was twenty-three years old and married, Whitaker began his service in the U.S. Army. He and his wife had two children at the time, a three-year-old son and a two-year-old daughter. Following World War II, Whitaker received a doctorate in optometry. He is an optometrist in Gering, Nebraska.

turn right around and send him back into combat time after time. He ended the war with us. He's still alive today. I hear from him through Christmas cards. I felt sorry for him because he was unable to really cope with combat.

Jay Dell Butler,[12] **Army infantryman, World War II**

Our two machine guns and the company I was attached to were assigned to walk right straight toward town without shooting back. The other two companies were to go around on the west side and walk through the town shooting from the hip at anything that moved. This was at eleven o'clock at night. So we started toward that town and got pretty close to it. Then the other companies started walking through the town, shooting from the hip. When you get every man shooting, it doesn't matter what he sees, he just shoots. If you see something, then of course you shoot, but the main idea is to shoot and walk. So that's what they did. They were lined up. I don't know how far apart they were; five to ten feet apart. When you start to shoot, then you start to holler and cuss. You kind of scream, yell. You could hear that roar of war. You just can't imagine that sound. Those men were walking and shooting. That's the way they took that town. Everything, including basements, was ripped up as they went through. They were throwing in grenades.

The Germans had about five or six men on the north side of town, trying to protect it. But we weren't shooting. We were trying to get in there with our machine guns. So we just walked, and they'd shoot at us, and we'd hit the ground and get up and go again, and they could hear us

---

[12] Jay Dell Butler was born 6 July 1923 in Tetonia, Idaho. After graduating from Teton High School and attending Brigham Young University and the LDS Business College, Butler began his World War II service when he was twenty-one years old and single. Butler has worked as a farmer, meat cutter, merchant, and Soil Conservation Service officer. He is presently a church custodian.

coming in the dark. So they'd just shoot in the dark toward us. This one time, I saw a fellow to my left, and I saw the flame of his rifle, and with the rest of them I jumped in the ditch to get out of the rain of the bullets. They were popping over our heads just like fire crackers.

As we got up to go, I looked to the right and saw another rifleman shooting at us, too. I could see the flame of his rifle. Just as I turned my head to the right, I got the bullet. I guess turning my head to the right saved me, because the bullet touched my jaw and it entered in through the top of my shoulder. The bullet went through me, but on its way, it broke my collar bone and top rib, and my rib punctured my left lung, and the bullet split the shoulder blade in my back.

So there I was in that ditch full of water. I couldn't get up. I don't know why, I guess it was a shock or something. The ditch water was ice cold, I knew that. Everybody had to go off and leave me. I said, "Lieutenant, I'm hit."

He says, "Ahuh."

I hollered at James Bradley and said, "I'm hit."

He says, "Ahuh."

But they had to leave me, and I understood that, so I didn't feel bad that they'd go off and leave me out there in the middle of the field, because I knew they had to keep going. So I was left out there alone in the dark, in that ice water, and I couldn't see anywhere. I heard one guy up the ditch a ways holler, "Medic," so that's what I did. Sure enough, four medics came up from behind us with stretchers. They knew we were getting into it, so they had the medics behind us, ready with stretchers. They were pretty scared boys. They weren't used to getting into battles. The guy up the ditch a ways, he was shot in the ankle. So the medics went up there to see him, and made sure he was all right, and then they came back and got me because I was hit worse.

They picked me up and put me on a stretcher and put me on their shoulders and started me back. They were going to take me to a M.A.S.H. [Mobile Army Surgical Hospital] unit. Just as I got up on their shoulders, the Germans threw in a mortar. It landed 150 feet from us. It scared them so bad that they all dropped to the ground and I rolled off the stretcher. It was kind of comical because I got back up on the stretcher myself. It seems to me that if I could get on the stretcher myself, I could've got out of that ditch. Then I noticed that I couldn't move my arm. I reached over, grabbed my sleeve, and pulled my arm up on me. I didn't pay a lot of attention to it.

They took me back to the unit where they cut my clothes off, and I suppose they figured I wasn't going to make it because I was shot up pretty bad, and they wouldn't take my boots off. My boots were full of that ice water, and my feet were freezing right off, but they wouldn't take my boots off, so I suppose that's one sign that they didn't think I'd live. Then a minister came up and had a word of prayer over me, and when I said, "Amen," he was kind of surprised.

Chris Velasquez, Navy combat photographer, Vietnam

The only wounds I ever received were from shrapnel. I was never hit by a round of ammunition. A round would tear a hole in a person. When I took a hit to my forearm I never even knew I was hit, I was so caught up in what I was doing. I took a few steps forward and noticed that my right side felt soaked. At first I thought I was perspiring, but then I felt my whole pant leg was soaked. I felt blood squishing in my boot and I thought I must have wet my pants. I didn't give it a second thought because there was a lot of that going on. I just kept doing my job.

The corpsmen saw me, rushed over and tried to get me to lay down so they could take care of me, and I said, "No, no, no, I'm all right." When I finally looked around, the muscle inside was starting to come out. It looked purple and ghastly. It looked like bubble gum! That's just exactly what the muscle looked like. They told me to lay down again, so this time I did. *Then* it started to hurt.

COMBAT ON THE GROUND

Lincoln R. Whitaker, Army infantryman, World War II

One time we were fighting fiercely to take the town of Krefeld, Germany. They had tank pits surrounding the town, so we couldn't get our tanks in there. It was up to the infantry to take it. We had to go through these mine fields with booby-trapped tank barricades. If a tank got close to these it would blow them up. We got through them and were crossing the field. At the end of the field were brick and stone courtyards. There was machine gun cross fire across that field from both corners. When we started moving across the field, we just ran as fast as we could. The men behind us were trying to knock out those machine guns with covering fire. They didn't succeed until we got into the town. My squad got right up next to the wall. We took a bayonet and put it on the end of a gun and poked a helmet up over that wall, and it got blown off immediately.

Our job was then to figure out how we were going to get over that wall and secure some part of that ground so the rest of the troops could come. We had to get in there and shut the machine guns off. Two of us would stand there and have a man run up and jump into our hands and we'd try to flip him over the wall before they could open fire. We got the first man over and we tossed his gun over to him. Got the second man over, and the third man and the fourth. We didn't have any casualties getting over that wall.

When we finally secured the town we found that twelve-, thirteen-, and fourteen-year-old German kids were guarding it. We took them prisoner. They were fanatical. They'd been brainwashed so terribly that they were absolutely fanatical.

That night, when we finally had a chance to regroup, secure our positions, and wait for daylight to take the rest of the town, I discovered that I'd had two bandoleers of ammunition shot off from my chest. I had bullet grooves through the front of my clothes and two buttons shot off. I also had four bullet holes in one pant leg of my fatigues. Yet I was untouched by any bullets. I felt very fortunate.

There were a lot of boys who found that they had wounds after we'd gotten in there and got settled down. One boy had been shot through the leg but wasn't even aware that he'd been shot until I noticed that his boot was bloody. We took his pant legs off and found that he'd been wounded. Another man had been shot through the lung and wasn't even aware that he'd been shot. He said he just had a little sting in his chest.

Dennis E. Holden,[13] Marine infantryman, Vietnam

One day I was on patrol, walking point[14] along this trail, with a man walking behind me called the "slack man." He's the one watching what was going on in case you missed something. As I was walking along I felt a thump: something hit me in the back and knocked me over. I reached around and felt my back to see if I'd been hit and there was blood all over me. What had hit me was the slack man's leg. He was seven or eight feet behind me and his leg had been blown off and had hit me in the back. This shook me up pretty badly. There was all kinds of shooting going on. There was screaming and yelling from all around. The shooting was so severe that I couldn't even get back seven feet to tie off my slack man's leg. He bled to death. As I was getting to my knees, a Viet Cong [VC] jumped out from beside the trail and smacked me in the face with the butt of his gun. Why he didn't shoot me, I'll never know. I rolled over, raised my M-1 rifle and emptied the clip into him. Today I wear a bridge where that VC knocked my teeth out.

---

[13] Dennis E. Holden was born 5 November 1947 in Latrobe, Pennsylvania. After graduating from Anaheim Union (California) High School, Holden went to Vietnam as a single nineteen-year-old. A life-long member of the Methodist church, Holden is the state education advisor for IBM in California.

[14] The point man is the lead man on a combat patrol.

George L. Adams,[15] Army wheeled-vehicle mechanic, Vietnam

While on perimeter guard we stood our duty in a bunker. The bunker was protected by fifty-five-gallon drums of a jelly-gasoline solution called "fou gas." The fifty-five-gallon drum sat on a one-pound stick of dynamite and had a white phosphorus grenade sitting on top of it. The "fou gas" was detonated by a Claymore mine charger. There were twelve Claymore anti-personnel mines on each side of the bunker, so there was quite a bit of armament used to protect the bunker. There was about forty-five feet of bare earth in front of the bunker, and then the first of five rows of concertina wire began. Concertina wire is a razor sharp coil of wire. Coils were stacked in rows, two coils on the ground and one on top. The rows of wire were spaced ten feet apart with trip flares on the ground underneath.

One night I was alone in the bunker. There was a requirement that two men be on duty at all times, but this particular night the individual who was assigned to me was out on patrol. A thick fog came in, and I couldn't see out to the first row of concertina. I was in the bunker for about five hours without anyone to talk to, not knowing what was in front of me, and being alone, I became really nervous. At about 4:30 or 5:00 a.m. I could hear something in the wires. I didn't know whether I was just hearing this or if there was really something out there. I left the bunker and moved out to the sides so that I could hear better. I definitely heard something moving through the concertina wire out in front, but I never was able to see what it was. At one point I heard what sounded like a kettle drum over in the direction of our "fou gas." It sounded like someone was try-

---

[15] George L. Adams was born 15 September 1949 in Provo, Utah. He graduated from Provo High School before going to Vietnam as a single nineteen-year-old. Currently, he is a service technician for Mountain Fuel Supply. He recently retired from the U.S. Army Reserve after having served for twenty-one years.

ing to take the white phosphorous grenade off the top of the "fou gas" container. I started yelling in that direction and giving orders like I was sending people in that direction. I heard some commotion and then nothing else; it was silent.

When daybreak came and the fog had lifted, the other individual returned from patrol and we went out to the area to see what had happened. We found signs that indicated someone had crawled through the full length of the concertina wire without setting off the trip flares. Someone had, in fact, been working on the grenade. We found where they'd moved the grenade. I guess my yelling and acting like there were other people with me scared the individual off. We found the areas where he came in and where he went out, but he didn't actually get anything. Out on the farthest part of our concertina wire, the trip flares had been disengaged. The flares were set up on a butterfly mechanism so that if the wire had been pulled too far or cut, it would've been tripped. They'd immobilized the butterfly and cut the wire so they could go through it. A lot of stories ran through my mind about people who went out on guard duty and the next day, when the second shift came out to relieve them, they were found with their throats cut.

David L. Evans, Army infantryman, World War II

One night we heard a group of German tanks moving around on the ridge that we were facing, and when the very first light of dawn finally started to show I looked out with my binoculars and I couldn't believe it. There were five German tanks sitting right out there on the ridge straight in front of me a quarter of a mile away. And then I realized what had happened. They'd been sent against our second battalion, which was on the far side of the ridge about a half a mile beyond. They'd apparently gone out during the night and gotten onto the ridge facing the battalion and then decided to spend the night and attack in the morning. They didn't know we were on the other side, so they just came over and the tank crews were spread around on the ground sleeping. The tanks were completely unmanned, just sitting there.

When I realized what the situation was, I decided that if we tried firing and getting the range, they'd just jump in the tanks and go and that would be the end of that. So I called around to all the headquarters and fire control centers I could and they coordinated fire from every cannon and all the howitzers that were attached to the division or part of the division that were in the area. We had about twelve 155-millimeter Long Tom guns that had been assigned to the division from army artillery, and they were all in the vicinity. So we calculated to have all the artillery that we could summon fire time-on-target air bursts over the tanks, first of all, to get rid of the tank crews. They fired three rounds, here and there, coming in different whistles. Then everything went off at once and just shook the whole hillside. Then the second round went off and then the third. By that time the tank crews were running in circles falling down all over the place as the rounds went off. The ones who survived ran over the ridge. They came under fire from the second battalion, which had been alerted. They finally took off along the ridge. Some of them got away. But we had set all five of the tanks on fire.

Lincoln R. Whitaker, Army infantryman, World War II

When we got to the aid station the nurses asked my companion if he was all right. He indicated that he was fine. His name was Scales. He was a little country boy from down in the southern states somewhere. I said to him, "It's going to be daylight before long. If we don't get back to the holes we won't get there today because we'll be in broad sight of the enemy." We got up to move and he said, "Boy, I'm stiff and sore." I said, "Let me look at you." I examined him and found he had a bullet hole through his back that he wasn't even aware of. I took him back into the aid station and told him to report there.

I found out later that this man sat all through the night at that aid station and never complained to the surgeons or the people in the hospital unit about his wound. Finally, after they'd taken care of all the wounded we'd brought in, one of the physicians walked out and said, "Can I help you?" He said, "I've got this little hurt in my back. I'd like you to take a look at it." It was a chest wound and the doctor said later that he was probably one of the worst wounded men there, but he didn't want to go ahead of anyone else. He wanted the rest of the men to be taken care of before he was, yet he had the most serious wound of any of them.

David L. Evans, Army infantryman, World War II
At night we had to go out and lie in the snow-covered fields outside the village so if the Germans attacked we could see them coming and be in a position to cut them down with automatic weapons and run back into the village. So we'd put on all the clothes we had—field jacket, with an overcoat across that, a little woolen cap under the helmet, and so on—and lie down on the sheet and pull it up around us so that we wouldn't stand out too much in the snow. Everybody was automatically on guard all night, every night. Then during the day they'd attack with tanks and blow the place apart. During the daytime they'd be attacking and at night we would be out there, so for seven days we hardly slept. That's the most exhausted I've ever been in my life.

Lincoln R. Whitaker, Army infantryman, World War II
The fighting was very fierce. We fought our way to within a short distance of Berlin. We then crossed the Ruhr River. When we got to the Ruhr River, the Germans were on the opposite shore defending their positions. We brought in enough artillery to lay down a barrage to cover every square yard from the far edge of the river for five hun-

dred yards deep with one 155-millimeter shell, and we thought no one could survive.

We eventually crossed the river in small boats, a squad at a time. We no sooner hit the water than the guns were roaring so loud that we couldn't hear ourselves think. We lost a lot of our good, strong combat men to nervous breakdowns at that point, and some of them were sent back. Some of them went berserk, and we had to hold them down and get them back out of the fighting so that we could continue on. We got across the river all right in my boat, but unfortunately a lot of them didn't make it.

We were to regroup on the far shore and march to a certain area where we were to meet. We were marching down a highway with hedgerows. It was dark, because we crossed the river before daylight. I was bringing up the rear guard with my squad. As we were marching I saw machine guns bristling out of these hedgerows that could've just cut us to pieces. I made my way to the head of the line and got hold of the company commander. I said, "Do you see what's in those hedgerows?"

He said, "Yes, keep quiet and keep moving."

Why the Germans never opened fire on us we'll never know because we could've been annihilated right there in just a couple of minutes. It's an eerie feeling to walk in front of the enemy when you can reach out and touch a machine gun on the nose and you know the enemy can start bursting those things any moment.

Danny L. Foote, Marine artillery, Vietnam

There were many times when we had people working with us who were Vietnamese, and then at night you'd kill somebody who was trying to penetrate your perimeter and it turned out to be somebody that you knew. So it was extremely frustrating to realize that you didn't really know who the enemy was. You felt extremely vulnerable. In a heated situation, your number one priority was to protect your life. You start thinking that you don't know who the enemy is, and you don't even know how old the enemy is, because I've seen Viet Cong ten and eleven years old, and I've seen Marines killed by nine- and ten-year-old kids. When you get in that atmosphere, your only instinct is to protect yourself at whatever cost.

Ninety to 95 percent of the fighting in Vietnam was done at night. That's something I don't think people realize. During the day our enemy blended in with townspeople, so it was foolish for him to try and do anything. The prime time for him to fight was at night.

From time to time we'd get ambushed, which usually meant sporadic sniper fire or machine guns set up to open up on you when you moved into a certain area. I'd say 98 percent of the time you never saw who was shooting at you. You just felt the effects of it. It would last for thirty seconds to a minute, and then it would be all over, and you'd be running around on adrenaline with nothing to shoot at. It was extremely frustrating.

Walter H. Speidel,[16] German Army Afrika Korps, World War II

Fighting a war in the desert was similar to fighting on the open sea. One night, after we had taken Tobruk and had set up camp for the night on our advance into Egypt, tanks and trucks suddenly rattled through the camp. Our guards had heard them coming, had even recognized them

---

[16]Walter H. Speidel was born 5 December 1922 in Stuttgart, Germany. Before entering the German army, Speidel completed his *gymnasium* education (thirteen years). He received his *Abitur* (diploma) after taking general education courses with an emphasis in languages and humanities. Speidel joined the army as a single eighteen-year-old. Since his military service, Speidel has been a radio and television station administrator, an account executive in advertising and public relations, an office manager of a hospital, and a translator. He also received a doctorate in German, and he is currently a professor emeritus of German at Brigham Young University.

as British vehicles, but had just let them drive through. Since we had never taken the time to put our insignia on the vehicles we had captured, nobody was ever quite sure if they were "friend" or "foe."

We had something of a gentlemen's agreement with the British not to start shooting unless we were in combat or if we were ordered to do so. For example, in battles when the front was moving back and forth, field hospitals would generally not be evacuated. The doctors and medics would continue their work, treating the wounded, whether German or British.

On at least one occasion I experienced this unofficial understanding. While in the desert near El Alamein trying to find and repair a broken cable, we were suprised by a British armored vehicle. The soldiers stopped, got out, waved, and came over to us. We talked for a while and then went our different ways. That was something I appreciated most and remember most vividly about the war in North Africa.

However, this could also be dangerous. During the final attack of the Allied forces on El Alamein in October/November 1942, a British "Dingo" shot up our "jeep" even though we had started to wave when we had seen it approaching. Apparently, they had not heard about our gentlemen's agreement. Luckily, only one of us was slightly injured, but we were stranded in the desert until a retreating German tank picked us up before nightfall.

Jerry L. Jensen,[17] Army Special Forces, Korea and Vietnam

They gave us an order to go out and neutralize a village. It was in an area that was strongly VC. The village

---

[17] Jerry L. Jensen was born 22 May 1932 in Olympia, Washington. Following his graduation from Twin City High School in Stanwood, Washington, and some university work, Jensen completed an eleven-month tour of duty in Korea. He received a bachelor's degree in chemistry from Washington State College

itself was considered to be Viet Cong. We had been through there several times and had had good experiences with the people. We didn't want to kill these people. The order didn't really say we had to kill them, it said "neutralize."

We went in and talked to the chief and told him what our orders were, that we were to neutralize this area, and asked him if he knew what it meant. He shook his head, yes, he knew what it meant.

I said, "Why don't we move you? We'll move you about fifteen kilometers?" I even had some helicopters brought in. We moved the whole village. After they pulled out, we torched the area. It was neutralized. That's how I handled that. I couldn't see killing these people. They were good people. They were very strongly involved politically and they were anti-Saigon, but they weren't VC.

Lincoln R. Whitaker, Army infantryman, World War II

A lot of our own artillery fired into us. As a result of our own artillery, many of our forward observers were killed. If the enemy could determine that there was a forward observer for artillery out there, they'd do everything they could to knock him out because they knew that the artillery then couldn't be directed towards them. We stood as much chance of being hit with artillery as the enemy did. I remember three instances in which we were caught in the midst of a rolling barrage. A rolling barrage is where they fire a barrage of ammunition in front of the infantry as they move. The forward observer keeps that artillery fire going

---

(now Washington State University) and a doctorate of divinity degree from Seattle Pacific College. A former Methodist, Jensen joined the LDS church in 1955. When Jensen left on his first tour of duty in Vietnam (1962-63), he was married and had two daughters (ages three and two). After his first tour, he completed two TDYs (temporary duty tour) in 1964 and 1965. Following his military service, Jensen did post-graduate work in counseling and guidance. He is currently an academic advisor and counselor and an instructor of religion and career education at BYU.

over our heads and lighting one hundred to two hundred yards in front of us, and then we'd move up behind it. If the forward observer gets killed, the orders are to keep moving. At that point you move right into your own artillery fire. We had a lot of men wounded in those types of situations.

Michael R. Johnson,[18] Marine infantryman, Vietnam

One of our guys who was a short-timer, about to rotate home, took a patrol out. (You only spent thirteen months in Vietnam, which was one of the major problems there, I think.) Since he was about ready to go home, he wasn't about to run a night ambush out where there might be some gooks. He was going to run a night ambush somewhere real safe where there'd be no chance of seeing anyone.

A lot of guys did that. They'd take a whole patrol of eight, twelve, fourteen men, lay them down and say, "If you make a noise, I'll shoot you. You don't make any noise. You lay there. You take care of the radio and if it talks to you, you give the required signals to tell them we're okay and we're in our area and everything is secure." The whole time they'd be maybe one hundred meters away from the base in the tallest, thickest brush they could find, hiding.

So this one night he ran us out to an area right in front of our base. It couldn't have been fifty meters away. We all got into this tall brush and lay down and he assigned, "You're awake. You're asleep. You're awake. You're asleep."

---

[18] Michael R. Johnson was born 7 January 1948 in Huntington, West Virginia. Johnson graduated from Huntington High School and attended Brigham Young University for one year. At eighteen he joined the Marine Corps during the Vietnam conflict. Johnson's tour of duty ended when he stepped on a booby trap and lost both legs. He returned home, continued his education, and became a junior high school science teacher. He currently teaches in Nikiski, Alaska.

He'd rotate every couple of hours. Around one o'clock the base got hit big. I mean the base really got hit. We could hear things going on around us while we hid in the bushes. The base was rocketed. Da Nang, which was near our base, was rocketed. They blew up our artillery, so our artillery couldn't take care of the rockets. The rockets could fire for a long time before we could get something trained on them to knock them out.

So here we are lying in front of our base, all this massive firepower pouring out into where the gooks were supposed to be, and they are firing at us because we are right in front. We were scared to death, and we couldn't do any shooting. We could hear the gooks running all around us, and I suspect they knew we were there and just avoided us because they figured we'd be killed along with them anyway. The big bases behind us fired up a bunch of big artillery flares that hang in the air for fifteen minutes from big, big parachutes. Those flares light up everything just like it's day. I mean it's so light you could read a letter. With twenty of those all around the air, it's just like daytime all around the base.

Here we are laying down in the weeds, and it's daytime all of a sudden. We were worrying about who was going to see us—the gooks or our hill? And how was our hill going to know who we were, since we weren't where we were supposed to be? We couldn't call our hill and say, "Hey, we've committed some kind of military treason because we aren't in the right spot." All we could do was lay there and keep dead quiet and hide—completely hide, I mean just make ourselves invisible if we could. It was a real tight situation in which we couldn't win. We couldn't shoot, couldn't run, couldn't do anything but lay there.

Calvin William Elton, Jr.,[19] Army Air Forces aircraft mechanic, World War II

We were at Nichols field in the Philippines at the time the war broke out. I was on the crew of a B-40 aircraft. We were digging a foxhole. We were parked next to a softball diamond and we were ordered the morning of 8 December 1941 to dig foxholes. We decided that we'd dig it over by the softball diamond because it was away from our aircraft. We first agreed, "Let's dig it right here on the pitcher's mound." Then we decided, "No, let's not do that, because it will ruin the diamond." We went over behind third base and dug it. Three of us dug a large foxhole that we could get into. At about eleven o'clock in the morning air raid sirens sounded, so we all jumped into the foxholes.

While we were still down in the foxhole we heard the bombs dropping. One dropped right near us. We donned gas masks because of our inexperience and everything. At that time I was nineteen years old, and the other fellows were just about the same age. We donned our gas attack gear. We had those on for about ten or fifteen minutes and we decided to take them off because everything had settled.

After about forty-five minutes, when the bombs stopped dropping and they sounded an all-clear, we crawled out of the foxhole. The foxhole was four to six feet deep and about six or eight feet long. Right on the pitcher's mound there was a big crater. A bomb had hit right there. All across the rice paddy you could see bomb craters. We were right in line with the bomb craters. We were between two of them.

---

[19] Calvin William Elton, Jr., was born 2 November 1922 in Dividend, Utah. At nineteen, after graduating from Payson High School, Elton joined the U.S. Army Air Corps, which became the U.S. Army Air Forces in 1941 before the United States entered World War II. Elton retired as a major from the U.S. Air Force.

Eugene E. Campbell,[20] chaplain, World War II

We were still heading north towards Berlin when I got orders to report to our headquarters in Fulda. I got to within ten miles of the city where a bridge was blown out. I talked to a farmer who was working in a field. He said there was a smaller bridge on a dam and he showed me where it was. My driver and I were all alone. We had a carbine with us, and that was the total amount of protection we had. We were under the impression that the whole region was under American control. Patton had sent his tanks down the main highway and taken the main cities, and then the infantry regiments like ours would come in and occupy the cities. He'd push on ahead and the infantry regiment would move forward on each side of his tank force and capture the little towns. That way we'd sweep across Germany. So the city of Fulda had been taken, and the Twenty-sixth Infantry Division occupied it. We were assigned to clean up the little towns on one side of the thrust. I crossed the bridge and came into a little town and I noticed that people were hanging out their sheets. I knew that this is the way Germans air their bedding. They do it almost every day. I heard some small arms fire, but it didn't seem like it was very close. We even had a flat tire in one town and we stopped and fixed it. We went through another town and finally drove into Fulda, but we couldn't find the Seventy-first Infantry

---

[20] Eugene E. Campbell was born 26 April 1915 in Tooele, Utah. He graduated from Tooele High School. After spending time at Snow Junior College in Ephraim, Utah, and graduating from the University of Utah with a bachelor's degree in history, Campbell married, had two children, and taught LDS seminary classes in Bicknell, Utah. When the United States became involved in World War II, Campbell volunteered to be a chaplain. Following his service, Campbell earned a doctorate in history, taught LDS institute classes at Utah State University, and became a professor of history at Brigham Young University. Campbell died in 1986.

Division. We found the Twenty-sixth Division Headquarters and finally got to talk to the Commanding Officer. He asked, "What are you doing in here?"

I answered, "I had orders to come here, as this is where our headquarters was supposed to be."

He said, "There's been a delay and they aren't in here yet. By the way, how did you get here?"

I showed him on the map and he laughed and said, "Congratulations, chaplain, you just conquered two towns." I didn't know it, of course, but they were surrendering when they were hanging out those white sheets. I thought they were just airing their bedding. I asked him what would be the best way to get to my headquarters and he said, "You need to go on the other side of the river." He gave me directions and we headed back. On the way back I passed our artillery outfit, which was setting up for a night attack. I saw German soldiers lying dead with their heads blown off. It was horrible. It was about ten kilometers back when I found one of our infantry divisions that was marching off for a night attack. I'd been all through that area and didn't even know that it was German territory.

Timothy Hoyt Bowers-Irons, chaplain, World War II and Korea

I remember one day close to the end of the war. Maintenance headquarters was up in a big old castle up in those tall mountains. I'd gone back to one of the other outfits and on the way back we saw some deer over in a field. I thought, well, fresh venison might be nice. So I said to Riley, my driver from Iowa, "Hold up a minute." He jumped out with his rifle and both of us felt a certain kind of feeling — I don't know why you get it, whether it's some secret sixth or seventh sense, but somehow you feel things aren't right. Maybe they are never right. Anyway, we took a couple of shots and I said, "Riley, let's get out of here." We got in that old outfit and away we went. We got up to headquarters, and about an hour later here came an armored car patrol that had picked up what was left of a whole company of

Germans. When I talked to them, I found out they'd been up in the timber in the hill right behind us trying to decide whether to shoot us or surrender to the two of us while we were trying to shoot those deer.

David R. Lyon,[21] Army artillery, World War II, Korea, and Vietnam

The achievements of which men are capable were evident at some of the bridges across a major river in northern Korea. Our Air Force took pictures of these spans with rails hanging down from the bank that they'd blown out with bombing attacks and were satisfied that no supplies were getting across the bridge. When we approached that bridge, I was in an advance party and they were still contesting it. We managed to survive only by being in the remnants of the only building in town with concrete walls, the vault area of a bank where large quantities of Korean currency were used to build fires and cook and to otherwise exist in the increasingly cold part of the year.

We were able to see how each night the North Koreans reconstructed the bridge, ran trains over it, and sent supplies on to the forces south of the bridge, and then apparently destroyed the bridge again before morning. This is something that boggles the mind. It was a manifestation of the enormous skill, dedication, and effort that can be brought to bear on tasks that are thought to be impossible.

---

[21] David R. Lyon was born 21 June 1920 in Salt Lake City, Utah. After graduating from Bingham High School, Lyon attended the University of Utah where he received a bachelor's degree in English. Soon thereafter Lyon received his Army commission; he saw active duty in both World War II and the Korean War, and as an officer he was involved in a support role during the Vietnam War. He and his wife have four sons (born in 1944, 1948, 1952, and 1956) and one daughter (1958). Lyon's last military assignment, from 1968 to 1972, was as the first professor of military science at Brigham Young University. He then served as director of community affairs at BYU before retiring.

Edmond S. Parkinson, Army Corps of Engineers, Vietnam

The old French colonial roads were narrow, curved, worn out, and caused considerable damage and congestion to Vietnamese commercial traffic as the war raged on. Our job was to eliminate the meandering of the road by constructing new cuts and fills, building new bridges, providing proper drainage, widening the traffic way, and paving the new road with asphalt.

Strangely, the Viet Cong seemed appreciative of our efforts in these upgrades and used the roads each night after we'd retreated into our secure base compounds. For this reason — the fact that we were perceived as contributing to the country rather than tearing it apart — the VC didn't particularly bother us; they seemed to appreciate what we were doing. It became a war of schedules. We, the U.S. troops, would use the road in the daytime, and the VC would use it at night. However, they had to let us know that they were always there and could wield their power whenever they desired. They harassed us with ambush attacks on the road periodically. We were fired upon in the supply and construction convoys we ran from base to base, and they probed our base camp perimeter defenses on a weekly basis with sapper attacks.[22]

Hyde L. Taylor, Army Airborne, Vietnam

I remember we were coming out of the mountains one time and were supposed to stop at a Special Forces camp that had a few rice paddies around it. We didn't work in open areas very much in the unit I was in. As we came out of the mountains and down into this open area, we thought for sure that the Special Forces camp would have this area all secured. We just let security go. We brought in all of our security and we were walking down a road just

---

[22] Sapper is a term used for a military engineer, usually someone who is trained to use explosives; a demolitions expert.

like farmers going to work. We walked right into an ambush. They ambushed us maybe two hundred yards from the camp.

The road was higher than the rice paddies. The Viet Cong were on one side of the road and everybody jumped off the other side. The radio operator and I jumped on the wrong side. We jumped on the side of the road that the fire was coming from. They had us pinned to the bank. I remember him cussing me out for making him jump on the wrong side of the road. It was almost in jest. Nobody was hurt and we were lucky. We were very close friends. Several times we were in that same situation. All of this firing was going on and here he was cussing me out for jumping on the wrong side of the road.

Douglas T. Hall,[23] Army Special Forces, Vietnam

As a kid I'd seen several war movies in which a soldier would say, "Here, sarge, take my stuff and this letter. I just know I'm going to get it in this next battle." And then he'd get killed. I thought, "Well, if I ever have a feeling like that about a mission, I'm not going to go." Near the end of my tour, we did get a mission, and I had a funny feeling. I thought, "I shouldn't go on this mission." I told this to Jim Vessells, my team sergeant, that I had this funny feeling. I said, "I promised myself that if I have a funny feeling, a negative feeling about a mission, I wouldn't go." Jim said, "I'm not going to make you go, I'll find somebody else." He was pretty understanding about it in the sense that although

---

[23] Douglas T. Hall was born 9 March 1947 in Payson, Utah. He attended Montrose (Colorado) High School and graduated from Orem (Utah) High School. Following a year at the College of Southern Utah (now Southern Utah State University) where he studied geology, Hall went to Vietnam as a single twenty-two-year-old. He is an attorney in Salt Lake City and has been a member of a Special Forces unit in the Utah National Guard since 1979.

he couldn't understand why I said that, he wasn't going to force me to go with him. So I sat and stewed upon it for about an hour; then I changed my mind and said, "Okay, I'll go."

Jim said, "Fine!"

We packed up all of our gear and headed off to Bu Dop, which was Jim's old Special Forces camp. From there we'd run our operation.

On that operation several things happened. The first thing was that we moved out to the river that separated Vietnam and Cambodia. At that river we sat up for the night with the idea of crossing at a fording area the next morning. When we got up the next morning, I tried to raise our air cover on the radio and I couldn't get anything. I checked out the radio and I couldn't make it work. We couldn't operate without the radio, so we decided we'd go back into Bu Dop. When we got back to Bu Dop and were checking out the radio, we discovered that the only thing that was wrong with it was that the antenna post was loose, which normally would've been the first thing that I would've checked out. But I didn't check it out that time. I tightened up the post and everything was fine. While we were sitting there getting ready to go back in to complete the mission, one of the camp spies came in. He reported that he'd crossed that same ford earlier that morning and had walked right into a North Vietnamese Army [NVA] patrol that was lying there in ambush. We knew that they would've been there when we'd planned to cross. Had the radio not caused us a problem, we would've walked right into them that morning.

If that didn't teach me, nothing would. So we turned around and made arrangements for the next day to go out with a company that was going on a sweep operation. We'd be the point element, and when we got to the border area we'd just slip away from them. While we were the point element and getting close to the river, the company ran into an NVA patrol and we got caught in the crossfire. That's where I got wounded. I suppose I'd call it a premonition that I ignored.

Dallis A. Christensen,[24] Army Airborne glider pilot, World War II

After we'd all landed, we tried to get the small groups together into large groups and to orient ourselves by maps and by the terrain of the ground. Then we headed for the area where our whole division was supposed to assemble. We had to fight our way there. We were surrounded by Germans and the Germans were surrounded by other Americans. We finally got to our assembly area by evening after we'd fought our way along all day. We established our unit, got our security out, and found out what the situation was like. I happened to be commanding a heavy weapons unit. We set it up and had a rifle platoon for security.

Unknown to us there was a German group in a barracks in a wooded area a couple of hundred yards away watching us dig in. They watched me kind of half-lying on the ground giving orders where I wanted the weapons and stuff put. They pretty much knew then that I was a leader, at least in the unit. One of them opened up on me with a Schmeizer machine pistol that shot about one thousand rounds per minute. It missed me by about six inches; it was just like a sewing machine going by. I was lucky.

---

[24]Dallis A. Christensen was born 25 August 1912 in Chester, Utah. After attending Ephraim (Utah) High School, Christensen graduated from Salina (Utah) High School. Schooling at Snow Junior College (Ephraim) and the University of Utah qualified Christensen for a bachelor's degree in political science and a teaching certificate before he entered military service at the age of thirty. He and his wife had a daughter while he prepared for combat at Fort Benning, Georgia. Following his World War II experience, Christensen worked for the Veterans Administration and then served as the executive director of the Central Utah chapter of the American Red Cross for seventeen years. Christensen died in 1987.

Jerry L. Jensen, Army Special Forces, Korea and Vietnam

We had two teams working in one area. This was down in the delta. In the delta you are mostly in swamp, with water up to your neck. It's pretty messy. We sent two teams in during this one operation. Charlie[25] had been in part of the delta. We were going in really to feel him out, to see what strength there was in that area. They put two teams together—two reinforced teams—so there were twenty-four troops. We went in from one area and the other team came in from their area. We met in a relatively dry area (relatively dry: if you lay in it you had to keep your head up).

There were some downed trees up ahead and as we approached them I rotated to point. We figured twenty minutes at point in the type of work we did seemed like a lifetime, and thirty minutes truly was; if you stayed at point for thirty minutes, your chances of being killed were probably around 80 or 90 percent. So we'd rotate about every twelve minutes.

I was at point, and finally we made contact with our other unit. I saw their point man. I signaled him to gather and meet. He came over. Of course, we'd gone forty to fifty meters ahead of our columns. We got out together and then we started moving over towards a little bit of high ground. It was out of the water and out of the muck. We started going over. I signaled him down and we started crawling in. We came up to some dead wood that had been knocked down or blown down. I started along this big, big log. I'm not quite sure what kind of wood it was, but it was hard wood. I was crawling along this thing, and I was going to look up to look over. To this day I remember there was just a voice that was loud as can be: "Don't put your head up. Flatten yourself." I did, I just flattened myself right in the muck. I figured it was the other man who said it. Just at that

---

[25] A nickname for the Viet Cong; also called Victor, Victor Charlie, Mr. Charles, and Chuck.

time a machine gun raked the top of that log. They'd watched us and they were expecting us to pop our heads up. Obviously it wasn't the other man who told me not to put my head up, because he caught a bullet right between the eyes. If he would've said, "Keep your head down," he wouldn't have stuck his head up himself. I asked my column, "Did any of you guys yell?"

They said, "No, all we heard was the machine gun."

Danny L. Foote, Marine artillery, Vietnam

I had one experience where I woke up in the middle of the night on the floor yelling, "Rockets!" and alerting my bunkmates that we were having incoming projectiles. We waited for something to happen, but nothing happened. Somebody said that Dan's just having a bad dream or something. Then all of a sudden we started getting all these rockets, and part of our hooch[26] was blown away. If we hadn't been on the ground when the rockets came in, we would've sustained casualties. Why I woke up when I did I really don't know.

Dallis A. Christensen, Army Airborne glider pilot, World War II

When we got our first taste of combat we were in a wooded area and I was company executive officer. I'd been given the assignment of taking the mortar sections of the company over to the edge of a grove of trees and setting them up. There was a foot of snow on the ground at that time. That was one of the worst winters Europe had had in twenty-five years, and it was somewhat foggy, so we couldn't see the Germans. I guess they could see us a little bit. Anyway they knew where we were and they were firing on us.

I and some of the men I was leading heard a bar-

---

[26] A hut or simple dwelling, generally referring to the houses of rural Vietnamese. However, as in this case, American servicemen also nicknamed their in-the-field dwellings "hooches."

rage of German shells coming in, which happened every half hour or so. I jumped into a little foxhole on one side of a log. Two or three of the men in the mortar section hit the ground on the other side. They didn't happen to hit a foxhole, so they just went down on the level ground. After the barrage was over, I got up and looked around. Just on the other side of the log, a shell had hit the tree above me. Because I was in the hole I was pretty well protected. I looked over on the other side of the log after the shelling had stopped, and there was one of our men. All there was left of him was just from the waist down; the top of him was gone.

Howard A. Christy, Marine infantryman, Vietnam

In March 1966, guerrilla warfare dominated the action in Vietnam. The Viet Cong operated throughout the country in small detachments carrying out raids of intimidation against the native populace and carrying out hit-and-run attacks and ambushes against U.S. and Republic of Vietnam [RVN] forces. It was still several months before the first large elements of the North Vietnamese Army crossed the Demilitarized Zone [DMZ].

The northern military zone was under the operational control of the Marine Corps, specifically the Third Marine Amphibious Force. The primary tactical element was the Third Marine Division, and I was the company commander of Company A of the First Battalion of the Ninth Marine Regiment of that division. Our primary mission was to root out the Viet Cong in the sectors assigned, then extend protection and assistance to the local people.

The Ninth Marines controlled a sector south of Da Nang, and the First Battalion controlled that part of the regimental sector that surrounded Hill 55, about fifteen miles southwest of Da Nang. The area was quite flat, dotted by numerous tiny agricultural villages separated by large rice and corn fields and coursed by several meandering rivers. Because of the flatness of the terrain, Hill 55 (fifty-five feet above sea level) somewhat dominated the surrounding area, and for this reason the battalion command post was located there.

On the morning of 21 March, Company A was given a mission of carrying out a "county fair" operation in one of the villages north of Hill 55. The village was quietly surrounded in the early morning darkness and at dawn troops entered the village in a "hammer and anvil" maneuver to trap any Viet Cong who might have been there (there were none), then waited for vehicles to arrive with food and medical support for the villagers. Two tanks and two amtracks[27] came into the village with supplies, and the villagers were invited to take the food and to receive medical care. We settled into the day-long routine of the county fair.

In the meantime, Company C of the same battalion moved to a position about two miles to the west across a river that determined the north-south boundary between our two companies. As the operations of the county fair were quiet and I had little to do, I listened in on the battalion tactical radio net for anything of interest. At about noon, Company C ran into an unusually large Viet Cong force and a substantial fire fight ensued between Company C, on the west bank of the river, and the Viet Cong force, estimated at company strength,[28] on the east bank. I alerted

---

[27] An amtrack is a small-armored amphibious vehicle used to transport troops. The name derives from "amphibious tractor."

[28] A *division* is the basic military unit that goes into combat. It is commanded by a major-general (a two-star general) and has 15,000 to 16,000 soliders. However, in some circumstances, divisions have had as many as 22,500 troops.

Generally, three *regiments* make up a division. A regiment is commanded by a full colonel and has 3,000 to 3,500 troops.

Three *battalions* make up a regiment. A battalion is commanded by a lieutenant colonel and has 850-1,000 soliders.

Three *companies* form a battalion. A captain commands a company and is responsible for 200 to 250 troops.

one of the platoons of my company to get ready to move and put the rest of the company on alert. A few minutes later the battalion commander ordered that one platoon be detached to battalion control and moved by helicopter to form a block position in the Company C area of operations. I detached the Third Platoon.

The Third Platoon was dropped in the vicinity of Le Son 5, a village on the east bank of the river on the west limit of my area of responsibility and just across from Company C's battle position. But rather than being placed where they might have been an effective block, the helicopters dropped the platoon into a dry rice paddy virtually on top of the Viet Cong company engaged with Company C. It was a hopeless situation. Many men were hit as they jumped from the helicopters. Seven were killed outright. The enemy had at least four machine guns, which raked the paddy with fire; all the marines could do was scramble for cover. Those not already killed or wounded jumped into bomb craters here and there and pandemonium reigned. It was impossible to get control of themselves let alone attempt to gain fire superiority; too many unit leaders (including the platoon commander) were hit by the initial fire and both the platoon's machine guns jammed and remained out of commission for the remainder of the battle.

Soon after the Third Platoon was detached, orders came to move the entire company into the fray at the river. We boarded the tanks and amtracks and headed due west to the river, where we jumped from the vehicles and quickly

---

There are three *platoons* in a company. A platoon is commanded by a lieutenant and consists of 35 to 45 soldiers.

Three or four *squads*, with 9 to 15 troops each, make up a platoon. A sergeant E-4, a "buck sergeant," commands a squad.

In addition to the above divisions, the Marines divide their squads into a unit known as a *fire team*. Each fire team has at least four troops: a fire team leader, an automatic rifleman, an assistant automatic rifleman, and a rifleman.

formed a skirmish line, two platoons abreast, the left platoon guiding on the river, and began to move north toward the firing. Within one hundred meters we met the lead element of a sizable VC force, those who had engaged with Company C. They'd apparently broken contact with Company C and, not anticipating the presence of any other marines in the vicinity, were withdrawing southward along the river bank—directly toward us. A desperate battle ensued at very close range, including hand-to-hand combat.

We didn't know where the Third Platoon was or how it was faring. But as the company moved north up the river, a faint cry came over the battalion tactical net. It was a badly wounded marine, a radioman, calling for help. The battalion commander personally came up on the radio and began to talk soothingly to the wounded marine: "Now son, hold on; we're coming to get you. Where are you?"

A faint reply, "I don't know. I'm all alone. Everybody is either dead or evacuated. Please come and help."

Eventually the battalion commander ascertained from the marine that a white-star-cluster signal flare was within his reach. The colonel explained how to employ it, and, at the count of three, the flare popped above me from a position only a few hundred meters away.

I told the colonel I had the flare in sight and that I was moving to that position with a small detachment. We found the wounded marine, alone and bleeding from a serious shoulder wound, lying among several dead troops. As there was no firing when we arrived, it was apparent that the enemy had left the field; indeed, they'd since met their fate upon running pell-mell into our skirmish line. We soon located two other pockets of live marines, still in their holes—one group of eleven men under the Third Platoon sergeant, and another group of five men.

I ordered every able-bodied troop to take firing positions and proceeded to call for a medevac [medical evacuation] helicopter for the wounded radioman. I needn't have called. A medevac crew had been in and out of that field all afternoon. In at least five sorties they'd already lifted out all the twenty or so wounded and were coming back for the dead. I helped the medevac corpsman put the radioman and some of the dead marines on the helicopter.

In about a half hour the helicopter returned for the last of the dead. Again I helped hand up the bodies, which we had to dump in a heap in the main compartment. Since there was no room left, the corpsman had to sit on top of the bodies. As the helicopter lifted off, our eyes met and I perceived the overwhelming sadness in him; he seemed to be in anguish and I was deeply moved by his compassion and sorrow.

By late afternoon all again was quiet. The company linked back up and we moved toward a nearby village with the intention of establishing our position for the night, sending out the last of the wounded, and getting resupplied. (The troops, having just experienced their first heavy contact with the VC, had rather recklessly fired almost all their ammunition.) But as we approached, we received small arms fire from the village. We went to the ground, returned fire, and I called for artillery.

The battalion commander, knowing what we'd already experienced, told me I could have all the fire support I wanted. I answered that I wanted it all. First came 105-millimeter howitzer fire, then 155-millimeter field-gun fire, then eight-inch howitzers. The rounds screamed in from every direction and blasted that little village. Then the guns ceased firing and two sorties of jet aircraft came in, one with napalm and the other with machine guns and rockets, which set what was left of the village on fire.

Then out of the smoke came a Vietnamese family. There was a grandmother, horridly burned. There was a mother, who was carrying a badly injured little boy about five years of age. And there was a girl about nine, apparently unhurt. They'd been caught, probably out working in the fields, by the barrage and couldn't get to the safety of the tunnels that honeycombed all the villages in the area. I ordered an immediate medevac for the injured grandmother and boy.

The mother placed the little boy in my arms—I don't know why—and as I held him he died. Shrapnel had penetrated his brain, judging from the deep hole in his forehead. The mother dropped to her knees, unhooked her long hair, and began to sway back and forth and wail in a manner that apparently is common for women in that culture. The little girl then went into a rage and beat her fists on me as I still held her dead brother. We stood there, stunned. All I could do was stammer that I was terribly sorry. Through an interpreter attached to the company, I tried to explain the previous battle and all the casualties we'd experienced—hoping somehow they'd understand and not hate me for what I'd done to them. It was pathetic. I think they could see how badly we felt, but they couldn't see why we had thrown all the firepower of a mighty army at their little village.

That night, after we'd moved to another village, set up defenses, and I'd written a report of the battle, I went to the back of the amtrack to get some rest. I lay down among the packs taken from the casualties and unknowingly put my face in the brain matter splattered on the pack of one of the dead. It was a fitting end to the events of an unforgettable day.

John C. Norton, Jr.,[29] Army infantryman, Vietnam

On 9 July 1972, I had my singularly most anxious experience. We'd been in heavy contact with the enemy that

---

[29] John C. Norton, Jr., was born 19 July 1947 in Fort Bragg, North Carolina. Norton attended Hampton Roads Academy in Newport News, Virginia, and graduated from Columbus (Georgia) High School before attending the United States Military Academy, where he earned a bachelor's degree in engineering. As a single twenty-four-year-old, Norton began his tour of duty in Vietnam. A member of the Episcopal church when he went to Viet-

day. Late in the day, the North Vietnamese opened up with a mortar barrage that I'll never forget. I was in a two-man foxhole. Fortunately, we'd cut some logs to put over top, so we had some overhead cover. I was down at one end of the hole and my counterpart, a South Vietnamese major, was at the other end of the foxhole. One of the mortar rounds came in and landed right on the corner of my position. When the round came in and landed and the explosion went off, our first reaction was, "This is it." As I was lifted up and then thrown back into the bottom of the hole, I thought to myself, "It's all over." As I began to regain consciousness, I thought to myself, "I wonder where I'm going to be when I open my eyes." I was totally amazed that I was still in the foxhole and still in one piece. My counterpart, as well, was kind of dazed. I got on the radio and got some air support to drop a few more bombs. The mortar fire quickly died down, but it looked as if we were going to be hit by a major attack that night. The reports were coming in from our security outposts that the enemy was moving in around us.

At that point I did something that I'd never done before. I found a little place there within our perimeter,[30] a little bushy area where I could be alone. I got down on one knee and I said, really in desperation, "God, if you get me out of this situation in one piece, I want you to know that I'm willing to dedicate my life to your work." And I thought to myself, "Gee whiz, that's a tall statement coming out of you." But I was totally sincere! And then I thought to

---

nam, Norton joined the LDS church in 1973. Norton is a lieutenant colonel in the U.S. Army. He has been a professor of military science at Brigham Young University since 1987.

[30] Or area of responsibility. During much of the Vietnam War, a perimeter was generally a 360-degree area around a military unit. In more conventional wars, such as World Wars I and II, a unit's perimeter was the area on its front, tying into other units on its left and right.

myself, "Well I believe in God, but what would that require of me?" I thought, "Well, would I become a Catholic priest, or a Baptist missionary, or an Army chaplain." I had no idea. Regardless, I got back in my hole that night and I felt better for having done that. Little did I know that a year later I would join the Mormon church. For the moment I was totally wrung out from the day's emotional drain. I leaned against the side of my foxhole and dozed off.

three
COMBAT IN THE AIR
★ ★ ★

Lawrence H. Johnson, Army Air Forces bomber pilot, World War II

It's interesting to see what happens to a group of people when they get into a combat situation. I guess in a way you could compare it to football, because if you don't stick together, you are going to lose the game. The crew members all depend on each other. If the engineer doesn't do his things right, the airplane won't fly. If the tail-gunner wasn't there, you'd be a sitting duck for any Zero[1] that came in on your tail, and if the navigator goofed up, he could cost you everything. I'd say there's a oneness you have together as a crew, and you know, when we started getting further along, some of our guys were developed way beyond their time because of the experiences we had. The squadron leaders would pull them off our crew and let them fill in for another airplane that needed a better bombardier or a better navigator, and you got so that was a major worry, because you were afraid you wouldn't get them back. The

---

[1] Nickname for a very fast Japanese fighter plane. It was a single-seated, machine-gun-armed, high-performance pursuit plane.

whole crew worried about it. Even if the gunner got sick, they were concerned because some other person was flying with them. It was interesting to see the sort of team spirit that developed. And nobody ever considered not doing their share or goofing off, because they just knew they were part of the group and everybody depended on each other. So even the ones that were scared to death every time they flew were there and prompt and ready and did their thing to the best of their ability.

Ted L. Weaver, Army Air Forces bomber pilot, World War II

Before we went to bed we didn't know who was flying the next day and who was not, or what aircraft we were going to be flying. In the morning they'd wake us and tell us that we were flying that day. Usually at two o'clock in the morning, we'd get dressed and shave and go to breakfast, which in our case was usually powdered milk, eggs, and toast. After breakfast we'd go to briefing, where they'd brief us on the weather for the day, what the target was, the approach, and what the intelligence had to report on antiaircraft fire and the possibility of meeting fighters. They'd also try to predict what kind of weather we'd have over England on our return. Then we'd go check out our parachutes. Our parachutes were ordinarily our own and kept in our own lockers. Each man always used the same parachute. It was fitted to him and kept in good order. They were periodically opened and repacked so that we'd feel like we could rely on them. We didn't wear the same clothing all the time. Some days we'd wear electrically heated flying-unit liners, with electrically heated gloves and liners in our boots. These plugged into a twenty-four-volt power supply in the plane. Other times we'd fly with just sheep skin trousers and a sheepskin coat. There was no heat in the planes. It would be ridiculous to even try to keep them tight enough to keep heat in them, because one shell would destroy your heating capabilities. The airplanes also weren't pressurized, so we all wore oxygen masks once we got above eight thousand feet. We'd also check out flak suits (bullet-proof vests) that would stop a piece of flak if it wasn't a direct hit.

During the first two or three missions that I flew with my own crew, my navigator and my bombardier, who had to stand on the flight deck most of the time, didn't check out a flak suit. They thought it was too bulky and hard to handle until the first time flak hit us and a piece of shrapnel came through the ship and tore some holes in it. After that, they were very happy to check out flak suits.

Ray T. Matheny,[2] Army Air Forces flight engineer and gunner, World War II

The flak just about unnerved me. Sam Henry, our bombadier, sang to us over the intercom, which helped relieve the tension. The lead ship in our formation went to full power to get us out of the line of fire, but the flak continued for such a long time. I heard a CRUMP, CRUMP sound and knew the flak was very close. Then to my astonishment I saw a hole appear in the right wing that pushed up a bright piece of metal contrasting with the dull-colored war painted metal on the surface. This wasn't only accompanied by a CRUMP, but the plane was lifted up several feet. More flak sound and holes appeared. Then I saw a red flash in the center of a flak burst. It was very close and there were those sounds again, not only of the shell exploding but of metal striking and tearing holes in my ship. For some reason I looked down at my feet, and there between my

---

[2] Ray T. Matheny was born 15 February 1925 in Los Angeles, California. Matheny graduated from John C. Fremont High School in Los Angeles. At eighteen he entered the U.S. Army Air Forces to prepare for his service in World War II. For Matheny, who joined the LDS church in 1951, aviation has been a career occupation. In addition, in 1968 Matheny received a doctorate in archeology; he is currently a professor of archeology and anthropology at Brigham Young University.

boots was a piece of spent flak! The fear I felt can't be expressed. I was so tense that for a few moments of this barrage of anti-aircraft shells I forgot the cold. I expected to feel, or maybe not to be able to feel, a searing, ragged piece of metal rip through my flesh to leave me bleeding helplessly and dying in the bitter cold.

J. Keith Melville, Army Air Forces bomber pilot, World War II

Bleckhammer was a tough target. The anti-aircraft fire was intense, it was a very long distance to fly, and if anything went wrong it would be difficult to get back to the base in Italy. Some of the targets deep in enemy territory were flown as shuttle missions from England, with the aircraft flying on to Russia after hitting the target, returning to their home base later. I didn't fly any shuttle missions. On our way to Bleckhammer we were heavily attacked with anti-aircraft fire (commonly called "ack-ack" or flak). Holes suddenly appeared in the plane as unseen flak tore through the aluminum sheeting or plexiglas. If you can see the smoke from the shell burst, you don't need to worry about it. It's when you can feel the burst lift the airplane that damage is being done. Our plane was being tossed around by burst after burst.

My bombardier, Jim Williams, was hit in the shoulder. Fortunately, the flak didn't sever any of his bone structure or enter his chest cavity, but it severely tore the muscles. It was good that we had an experienced navigator with us, because he knew just what to do to take care of our injured bombardier. Both of our waist gunners were hit with spent flak, Bill Lazar in the shoulder and Berlin Runyon on the forehead. Neither one was seriously hurt. Runyon wasn't wearing his helmet on this flight—but he did on every subsequent mission!

The group successfully dropped its bombs on the ball bearing plant and rallied off the target. I was flying in the seventh position of my squadron, which was flying in the fourth position of the group. This was the worst position in the whole group, but the youngest pilot usually doesn't get the better slots. The plane had one engine damaged, which was smoking badly, but I didn't turn it off and feather the prop[3] because it was better on fuel consumption to run on four engines than on three. It was a long way home. The copilot continually asked me on the way home if we should feather the prop, but I said: "No, let's keep it running as long as we can."

We knew there may be problems on landing. We'd been sent to the Foggia main airfield to facilitate getting Jim to the hospital. The control tower, on learning that in addition to the smoking engine our hydraulic system wasn't operating, directed us to land on a dirt strip parallel with the runway. I asked Jack McCoy, my copilot, to pump the hydraulic system manually, and the gauge registered that we had pressure for the brakes. Jack told the crew to brace themselves. When I applied the brakes, I found we had brakes only on one side. This turned us in the direction of a row of British bombers being prepared for their night missions. I straightened up the plane with the two engines on the right wing, one of which was the smoking engine. We were almost to the end of the dirt strip, and not knowing what the field beyond was like, I asked Jack to unlock the tail wheel. I applied the good brakes and groundlooped the plane. My crew chief, Carlos Verduzco, said he watched the left wing tip cut part of a circle in the dirt as we went around. We were all thankful to be on the ground alive!

---

[3] To "feather the prop" is to turn the propeller blade into the wind so that the propeller shuts down without turning the engine off.

Ray T. Matheny, Army Air Forces flight engineer and gunner, World War II

I was awakened with a start. "The jeep, the damn jeep is coming" went through my mind and I hoped it was a dream, but the tire-crunching sounds in the heavy frost continued. I heard the orderly walk over to the barracks next to ours and after about five minutes his footsteps came to our porch. I dug my fingernails into my palms and silently cursed. I heard sounds from others indicating that they were awake also. Yes, another mission call at 02:00 hours.

It was to be the same mission we'd aborted the day before: Bremen, one of the best-defended targets in Europe. I wondered what was so important about Bremen that we had to go back so often. It seemed like a mistake to send us out again on the twenty-ninth when the mission had failed the previous day. From what I read in the British intelligence reports, the Germans would know all about the aborted mission and would be waiting for us. The briefing officer gave us the same target areas along the docks at Bremen. He warned, however, that today would be cold—colder than ever recorded before, down to fifty-five degrees below zero centigrade. We'd have to keep circulating the oil in the propellers, and he emphasized that we couldn't touch any metal with bare hands nor let our faces come in contact with anything.

We were issued our permanent parachutes and custom-fitted harnesses. I received a chest pack, as it was impossible to wear a regular parachute in the upper turret. The ball turret operator, Tex, also got a chest pack.

Dunning, the pilot, gave me the bad news. We'd been assigned to the oldest B-17F in the 379th. It had been a general's plane that had extra armor plating installed and "bullet-proof" one-and-a-half-inch glass in the cockpit windscreen and side windows. But the plane was old and worn in every way; even the engines had more hours on them than usual, and I noticed severe oil leaks on numbers one and two. Missing were the wide-blade propellers that our "G" model had to provide greater thrust at takeoff. This "F" model had "speed props" that gave them an advantage in cruise configuration, but there had been complaints about takeoff performance with these propellers.

It was true. The airplane gained speed so slowly for takeoff with combat loading that if it weren't for the dip at the end of the runway it wouldn't have made it. We were required to carry sixteen bombs that weighed five hundred pounds each, a full load of fuel, ammunition for twelve guns, and emergency gear. Despite the fact that the "F" model didn't have a chin turret and carried one less gun, it weighed more than the "G." Apparently electronic gear had been installed for the general that required extra wiring, mounts, and power supply, all of which were still in place. Nelsen, the right waist-gunner, discovered extra armor plate in the waist gunner positions, and this airplane's gross weight was over what the manual said it should be.

The weather was cold and clear, and it seemed that we could see across the continent. Everyone but me test fired their guns over the sea at low altitude. I was thinking about the forecast of intense cold, and I worried about my guns malfunctioning. I'd repaired my electrically heated undersuit by rigging up a light bulb with wires, a set of flashlight batteries, and a needle probe. I'd find the broken wires by checking for continuity, and where a break was found I spliced the wires. I also made up a small set of "safety" wires and attached one to one of the two electrical plugs for each glove and sock. The gloves were particularly bothersome and seemed to fail more frequently than other parts. If a glove or a sock failed, the entire suit would shut down due to a series circuit. My safety wires could be used to short a defective glove or sock and still make the rest of the suit provide heat. Of course, the glove would be cold, but you could at least stick your hand into a warm place from time to time. I'd repaired all of the suits for our crew and provided them with short-out wires.

After three and one-half hours or so we were at twenty-five thousand feet near the coast of Germany and we finally saw a few of our P-47 fighter escorts contrailing above us, but they only circled once and headed home. Their range with an auxiliary seventy-five-gallon external fuel tank was only 340 miles. We entered the coast south of the Weser River this time and climbed to our assigned altitude of twenty-eight thousand feet. By now the cold had reached the lowest reading on the free-air temperature gauge: minus sixty degrees centigrade. It was necessary to cycle the propellers every fifteen minutes to keep the oil inside the propeller domes from turning to sludge that could quickly cause a runaway engine condition. The inside of my turret had accumulated a layer of ice from the exhaust of the oxygen mask that rose as a misty cloud and turned to ice as it contacted the metal and plexiglass dome of the gun turret. The ice began at the top and had now worked downward to where there was only about two inches of ice-free visibility at the bottom of the plexiglass.

There was a fighter attack, but the German fighters made only a single pass and then disappeared. I heard guns fire only a few rounds, then silence. Gunners reported that their guns failed to operate. More fighters were reported, and I managed to sight one through the icy coating of the plexiglass and to fire a burst. Obviously my guns were working well. I knew what had happened to the other guns. They were plugged with congealed oil, which I'd carefully wiped off of my own guns, and the test firing at lower altitudes caused condensation to form ice outside the guns. The German Me-109 fighters had the same problem, which is why they could only make one pass, fire their guns only a couple of rounds, and then head back to the airfields down below.

The cockpit was also beginning to ice up and Dunning called for the ice scraper. I handed it to him, but the ice was so hard he couldn't remove it. Visibility in the cockpit was becoming seriously impaired, and Dunning and copilot Harper were flying by watching the ghostly images of the planes near us. To make matters worse, we were flying in the lowest position in the entire group of thirty-four airplanes, a spot affectionately called "Purple Heart corner." The planes were stacked in three squadrons of twelve, although we were short at least two planes that day. Each squadron flew in a position higher and forward of the one below so that when we were under fighter attack we could tighten up for maximum concentration of defensive firepower. During the bombing run, we loosened up the formations so the bombs dropped from the planes above wouldn't strike the planes below. The problem now was that the interior of the bombardier's and navigator's compartment had iced over and they could no longer see the actions of the lead ship.

Soon the flak became an intense barrage that laid down what we called a "blanket" of black puffs of smoke and destruction. The pilots complained that they couldn't maintain a very close formation, and they depended on me to tell them when we got too close to another plane. The bottom of the plexiglass in my turret remained clear, but only in a narrow horizontal band approximately one to one and one-half inches wide. From this restricted view I guided the bomber. I saw that other ships were beginning to have similar problems, but because of the thick glass in our ship, we experienced it long before others.

I saw the bomb doors of the other ships open. I called to Henry, the bombadier, and gave him the signal. He asked me to give him the cue when the bombs were away. The flak was heard again and I thought I heard the metal strike. It was a nervous few moments as I tried to guide Dunning, the pilot, in a formation with only a small view provided by the ice-free band at the bottom of the plexiglass. I saw the bomb doors close on the other ships. "Damn, damn," I said over the intercom, "we missed the bomb run and we are going to make another pass."

The pressure to keep us from colliding and the

knowledge that we'd be exposed to the "flak run" again were plying heavily upon me. As soon as our huge formation swung around north the flak tapered off. In a few minutes Nelsen called in two *schwarmen* of Me-109s bearing on our tail. Since all of the guns on our ship were out except mine, I swung the turret in the direction of the attack. We knew that the fighters would only be able to make one pass, but that one pass could be fatal for a crew member, an entire crew, or a ship. I soon picked up two Me-109s and tracked them as best I could, but the turret moved in a funny, buffeting way. I got in two short bursts with the .50-caliber machine guns and the targets moved out of my field of vision.

I could see in the slit of vision in my turret another group of bombers coming from the coast that was timed to follow us to the target. Now these two groups would be closer, and maybe the effect of bombing more severe on the group below. I worried about our group turning fast enough to cycle in with the stream of incoming bombers – especially now, as apparently nearly everyone was having some visibility problems. We made the turn and I could see the magnificence of the other group, so close but slightly lower with the sun glistening on the plexiglass of the nose sections and with the streaming clouds of water condensation trails that the ships generated. The formation looked like a giant swarm of insects in slow motion. But I was soon jarred out of my reflections by a glimpse of the shadow of a wing over my turret. I frantically told Dunning, "Let her down, back off!" I saw through the little slit of the frosted turret a wing tip of the ship in front and to the right. Nelsen said it was close: there was no more than a foot between the wing and our propellers!

The flak began again, but it seemed less intense than before. I thought that perhaps the other group close behind us was drawing some angular fire. "Those damn 236 flak towers down there and the immunity of their shells to the cold," I murmured to myself. The bomb doors came open again, and finally the bombs were away. "Head for home," I thought. "Let's get the hell out of here."

After what seemed a long time we passed the flak zone on our run for home. Dunning said that he was full out on power and I could see that we couldn't keep up with the formation. Also, a white coating of ice had now formed over the entire cockpit, giving it an eerie appearance. No instruments could be read because of ice on the gauges, and Dunning required a reference to the horizon. He said that it was getting more difficult to fly. I suggested the autopilot, but apparently the hydraulic fluid in it or some of the lubricants wouldn't allow it to work properly. I got out of my turret and tried to open the side window, which Dunning and Harper had tried before, but to no avail. I suggested the fire axe and Dunning agreed. Try as I might, the axe wouldn't break the window because of its great thickness and temper. I took a heavy swing with the sharp pick end opposite the blade and it imbedded into the glass. Seeing a new possibility, I swung again, only harder, and got a good puncture. Now that the pick end penetrated sufficiently, I pulled hard toward me and pushed against Dunning's seat with my feet. Slowly the window slid back a fraction of an inch. I repeated the process two more times and finally got the window slid back enough to get my gloved fingers on the glass and draw the window back. Dunning could finally see out this small window.

The left waist-gunner called out fighters and they took a shot or two. This time the fighters didn't dive back home but slowly drew up on us from behind, getting closer and closer. We'd now fallen back from the formation by a quarter of a mile and two Me-109s flew off our left wing a few hundred yards away. The left waist-gunner managed to get a single shot, then hand-charged his gun and got another. Then his gun quit. I watched as the two 109s moved closer until they were a scant forty to fifty feet off our left wing, flying in formation with us! I tried to bring my turret around, but the hydraulic valve just kept popping off and producing a pounding sound. The turret moved very slowly, chattering with the frozen grease in the track, but it never came around for me to line up and blast those Me-109s out of the sky. My guns were the only ones working and I couldn't shoot!

The graceful lines of the Me-109s became apparent with their landing gear and flaps up, lazily cruising with us. The pilot of the closest one waved in a friendly way to us, and I responded by raising my middle finger to him. He was no friend. Perhaps he'd killed some of my friends or would live to kill me. I could see so clearly the little swastika on the tail and the iron cross, which was partially obscured by what looked like "25." They soon departed and we were left to catch up with the group, now about a mile ahead of us.

I suddenly noticed on top of the cowling of number two engine what appeared to be a dirty rag. I stared for a moment before realizing that a rag couldn't be lying on top of the cowling while we were going nearly two hundred miles per hour. I thought that something was wrong with my mind as I watched the rag grow in size. Then it dawned on me that this was engine oil spewing out of the breather that vented at the top of the cowling. The engine had failed and was pumping hot oil into the atmosphere, but it was so cold that the oil congealed instantly in a blob on the cowling, looking at first glance much like a rumpled rag. I told Dunning of the situation and said that we must feather the propeller or soon run out of oil, at which point we'd be unable to feather it. Harper objected to shutting down the engine and argued that fighters would get us, but with the situation of the day I countered this. With the propeller feathered the old "F" slowed down and Dunning decided that soon we'd have to try and get back alone. I noticed the throttles were white with ice, not from the exhaust of the oxygen but from the perspiration of Dunning's right hand. He wore only a leather glove to operate the throttles and I'd shorted the glove plug of his electric suit so the rest of the suit would heat. The group leader finally slowed down as we passed through twenty-six thousand feet.

It was then that I noticed a German Heinkel-177 bomber off the right wing flying even with us but just out of the range of our guns. I could see a gun firing slowly out of its left fuselage gun station. Occasionally I could see a faint white arching trail of smoke as shells from a cannon were lobbed into our formation. These were thirty or forty-millimeter cannon shells, which could severely damage a B-17. The really disturbing factor was that we couldn't do anything about the Heinkel-177 because it was flying out of range of our guns.

We fell steadily back from the group and finally Dunning decided to drop down to an altitude where the old plane could operate better on its three remaining engines. Dunning leveled off at about eighteen or twenty thousand feet to cruise homeward. The navigator complained that he still couldn't read the instruments. Harper realized that none could be read in the cockpit either. We didn't know our directions well and had now dropped between cloud layers. I tried to clean off the ice from the magnetic compass, but nothing could be done with it. Dunning stuck his head out the window to see if a hole in the clouds could be found. The slip-stream promptly blew off his goggles, so I gave him my blue-tinted ones. He could see nothing but clouds. I suggested that we drop down lower and melt the ice on our instruments. Dunning let the ship down below the cloud layer, risking flak and fighters, but it was imperative to find our way. The descent through the clouds was perilous enough, but without instruments of any kind, it was doubly so. Dunning was nervous, and I just stared out the small opening I'd made by sliding the cockpit window back. Harper stared down at the control column but didn't touch it. It was a great moment of relief when Dunning spotted the ground at about twelve thousand feet. It was warmer and the ice inside the cockpit began to soften, but it still covered the instruments.

It was hazy and dark, and Dunning couldn't recognize any landmarks. I managed to scrape the ice off the magnetic compass and saw that we were flying on a 120 degree track, which would've led us a long way off course. Dunning swung the ship around to 270 degrees and thought that we at least ought to see some landmarks, especially near the coast.

At this point I was concerned about fuel consumption, as we had full power on four engines for a long time and then full power on three engines after number two had failed. This meant that fuel must be low, even if we still couldn't read the instruments.

I adjusted the engine power to eighteen hundred RPM (revolutions per minute), fuel mixture control to autolean, twenty-nine inches manifold pressure, cowl flaps closed. Our airspeed dropped to 135 miles per hour, but it should've been ten miles per hour faster. We needed to maintain as much altitude as possible for two reasons: first, we were flying over enemy land all the way to the French coast, subject to murderous ground fire, and almost any German airplane could attack us at our slow speed; and second, the airplane was overweight and performing poorly on three engines, and if another engine quit the old ship would go down. We continued to fly in clouds to avoid being seen from the ground. Soon we were flying between cloud layers, and that made us feel more secure.

The frost slowly melted on the instrument panel and one by one, the glass faces on the gauges were cleaned by our gloved fingers so we could see what the instruments had to say. The three engines showed normal readings on all the instruments. The only disturbing reading was the fuel. The fuel in the three tanks was low. Number two engine's fuel — over two hundred gallons — was trapped by a frozen fuel transfer valve, which I had unsuccessfully tried to open.

We slowly lost altitude at the rate of about one hundred to two hundred feet per minute. The fuel gauges were reading very low. We were flying between cloud layers and had no idea where we were. Dunning was simply flying due west on the magnetic compass—blindly guided, as it were, without a view of the ground or sea that might lie below.

Suddenly the warning light on number three engine's fuel tanks flashed on, giving us about ten minutes of power before that engine would quit. Henry then excitedly cried, "There's a hole in the clouds ahead!" Then he said, "There's the coast of England!"

We all saw the small hole in the clouds that gave us a brief glimpse of the sea pounding on the rocks from our perch five thousand feet above. Dunning called for full flaps and landing gear down and began spiraling tightly through the aperture until we broke out of the overcast at about eight hundred feet. I called Dunning's attention to the fuel warning light that had just flashed on for number one engine's fuel tanks.

Dunning said, "Matheny, get the crew ready for a crash landing."

I dashed back to the radio compartment and got the crew members to sit against the bulkheads, draw their knees up and place their heads down and forward.

I got back in time to see Harper pointing at a patch of ground off the right wing. "It's an airfield," he shouted in the noisy cockpit, where both side windows were open. Dunning banked the ship for a better look when the warning light for number four engine flashed on. I thought the number three engine was due to quit in about a minute or two, and I kept my eye on the fuel pressure gauge on the right instrument panel. Dunning lined up on the airfield, but I couldn't see it. What I saw was a small dirt strip, about five hundred feet long, and an area at the end of it cleared of trees, but with hundreds of stumps sticking menacingly out of the ground. There was no time to check further. Dunning was trying to land, but there was a tail wind. Dunning badly overshot the small strip of ground and opened the three throttles full, but nothing happened. The propellers had been set at eighteen hundred RPM and full open throttles brought no additional power at sea level. I instantly brought the propellers up to full RPM and the turbosupercharger controls to forty-six inches. We weren't fifty feet over the trees when the ship responded. Dunning brought her around at about 105 miles per hour. Dunning and Harper seemed to be struggling with the ship to keep it from stalling. The power held as the pilots made a base leg turn, while the fuel pressure went to zero on number three engine. Dunning and Harper got the ship lined up and the wheels touched down on the end of the small dirt strip. The big ship rolled on the main gear wheels at full brake, then skidded off the cleared land and on through the tree stumps, across a ditch, then another, finally coming to rest after slicing through a heavy wire fence. At that point the tail wheel hit the ground with a final crash.

Ted L. Weaver, Army Air Forces bomber pilot, World War II

Once we went on a bombing mission over Berlin. We'd spent nine-and-a-half hours in the air. We'd gone clear up to the North Sea and had come back over the northern part into Germany and flew right over Berlin. That was a sight I'll never forget. We were flying at twenty-five thousand feet and the entire city of Berlin was invisible because of puffs of black smoke from anti-aircraft guns. There was a curtain thousands of feet deep over that whole area of Berlin. They were throwing up a blanket made from the shells exploding.

We made it through the mission and went back to England. Our bomb load that day was fifty-two one hundred-pound bombs. We didn't have that many bomb stations. To hang that many bombs, we'd have three bombs on a bomb station, with two of them hung over the top of the other one with cables. They'd release three at a time from a station. Sometimes they'd jam up and pile up in the bay if they didn't go out just right. This trip we figured we'd gotten them all out. No one had gone back into the bomb bay to really check to make sure, because it felt like we'd unloaded. Usually when you unload, your ship would take a definite rise.

When we got back to England, the squadron leader chose to go over to the west coast and drop down under the overcast and fly back to the base across England at five hundred feet. There were twelve bombers flying in formation, three directly below the lead ship, three down below them, and three above the lead. That spread us far enough to where the top three were flying in the clouds and the bottom three were jumping tree tops. That was how we were working it out. As we got a couple of minutes away from the base, my engineer tapped me on the shoulder again and said, "Skipper, you better ask permission to break away and land first if you can. We are low on fuel."

I said, "How much fuel have we got?"

He said, "Oh, about three tablespoons in each tank." (He was kidding).

I asked the navigator what our ETA (estimated time of arrival) back to the field was, and he said, "A minute and a half." So I called the lead ship and received permission to break away and land ahead of the rest of the ships because of low fuel.

As we went over the field for the initial pass, I gave the break-away signal to notify my two wing men that I was breaking out of formation. That signal was three sharp dips of the nose. I made the first dip and as I came down on the second dip there was a thud on the plane. I went on and broke out of formation and swung around into a right hand pattern. My engineer came and tapped me on the shoulder again and said, "Skipper, you better make that a close base. Don't spread it out or you are going to run out of gas before you can get her down." So I cut the pattern short, made a real close pattern and steep approach, and put it down on the runway. As I rolled to the end of the runway, my two outboard engines cut out for lack of fuel. My parking area was just off to the right of the end of the runway and about one hundred yards back and into a parking circle. I coasted as best I could, doing as little braking as I could get away with, and got around into my parking stall. As I spun it around to park, the other two engines quit. I was completely out of fuel. I'd hardly gotten the brakes set when the engine switches all shut off.

I'd just started climbing out of my pilot seat when I saw the commanding officer's jeep come tearing into my parking area. We climbed down out of the bomb bay, and by then he'd pulled up to a halt and jumped out of his jeep and was running toward me. I called the crew to attention as they came out of the plane. I saluted the commanding officer. He didn't wait for me to ask. He said, "What in the blazes are you doing taxiing down the runway with one of your crew members sitting on the bomb bay track cat walk?"

I turned to the crew and asked them, "What's he talking about?"

My engineer stepped forward and said, "Well, Skipper, I didn't want to worry you when you were coming in to land. Remember when you made that dip over the front of the field?"

I said, "Yes."

He said, "We had a bomb go out and it took the door off." (The bomb bay doors were closed for landing.) He said, "So I got out on the cat walk and held the door up while you were taxiing so it wouldn't drag." The commanding officer and I didn't have much to say. I said, "Well, I guess all we can do is wait until we hear where that bomb landed. We'll hear, I'm sure."

Fortunately, it was still unarmed. Even though it had broken off of the rack, it hadn't fallen enough distance to spin the arming vein off—this takes a couple thousand feet. So it was still safe. Three days later the base got an irate call from one of the local English ladies who had gone out to hang her clothes and had found this bomb between the house and the clothesline.

Dallis A. Christensen, Army Airborne glider pilot, World War II

We loaded into gliders and C-47s to make an airborne drop across the Rhine River. I went across that morning in a CG4A glider. CG4A gliders had only a doped canvas skin that was strong and stretched over metal frames. In my glider I had a jeep, first aid supplies, a machine gun, machine gun ammunition, and one machine gun crew, plus a pilot and a copilot. There would be two gliders at the end of three hundred feet of nylon cable attached to the C-47 tug ship. When we got to where we thought our drop zone was, the pilot would just reach up and hit a little release and we'd be on our own. We were apprehensive about whether or not we were going to reach the ground safely and about whether it was going to be a good landing or a bad landing when we made it to the ground.

When we got to where we were ready to drop, the Germans had moved in some batteries of anti-aircraft guns. They knew we were coming. The anti-aircraft shells would explode in the air, and it sounded like we were on the inside of a popcorn popper, the flak coming through the canvas.

After we reached the ground in our glider, or parachute, or whatever, we never knew if we were going to be where we could fight our way out or if we were going to be surrounded by the enemy. Fortunately, when our particular glider landed, we were able to get out. We were a little ways away from the enemy, so we were able to get our machine guns, ammunition, jeep, and supplies out of the glider. Then we moved up a large canal that went through the drop zone where we landed. We got up to the bank of the canal and saw men dropping in parachutes and gliders, some coming down in flames. We saw one glider land within fifty yards from us that was on fire before it ever hit the ground. When it hit the ground it just disintegrated. Of course, it was full of men.

David I. Folkman, Jr.,[4] Air Force fighter pilot, Vietnam

After I'd been flying for six months, I had a mission to fly up to the Cambodian border where the Viet Cong had a camp. There was a truck they were using, a yellow truck. Our mission was to bomb that truck. My wingman and I flew in there with napalm to drop on this truck. Evidently it was out of gas or something, because they left it on the road. My wingman made four passes and I made three. We missed the truck all seven times because the trees were about two hundred feet high and the truck was on a narrow road. I was determined to get the truck, so I went below the tree level and dropped my bomb. I got the truck, but when I tried to get out, I saw the road had turned a little bit. I was so far below the trees that I didn't have room to get out. I crashed into the top of a tree.

The airplane shook and I closed my eyes. I thought I was dead. A second later it was quiet and the airplane was still flying. I opened my eyes and saw I was about twenty feet above the trees, so I pushed the power up higher. I looked out the right side. Half of my wing was gone, cut right in half. My gas was flowing out and I knew there was no way I was going to make it home. I climbed for altitude and got up to about ten thousand feet. I contacted my wingman and he called for emergency rescue. I had to get out,

---

[4]David I. Folkman was born 9 March 1929 in Elyria, Ohio. He graduated from Ogden (Utah) High School. Folkman earned an associate's degree in civil engineering from Weber Junior College (now Weber State University). He then attended Brigham Young University and the University of Utah, earning bachelor's, master's, and doctor's degrees in history. When he was forty years old, married, and had three daughters (ages fifteen, eleven, and seven) and two sons (ages ten and five), he left his teaching position at the Air Force Academy to serve in Vietnam. After retiring as a major from the Air Force, Folkman worked as a real estate developer for eighteen years. He now teaches history at the Utah Valley Community College in Orem, Utah.

so I selected an area that was clear of trees. I then said a prayer. I reached down and pulled up on the handles. I was out of the airplane and starting to float down. It took me about fifteen minutes to descend because I'd been up there so high. I tried very hard to stay in the center of that meadow, but the closer I got, the more the wind blew me towards the trees. I landed on a fallen tree on the edge of the meadow. I hit it just below my hip and it wrapped me around and slammed me against the ground. It knocked all of my breath out; I couldn't breathe. After I regained my breath, I got on my knees and started to take off my parachute. I heard a helicopter, and as I was unbuckling my parachute, two guys on both sides of me grabbed me and ripped off my parachute. They then grabbed me and half carried me to the helicopter. They threw me in the back, and as we were going up I saw in the woods that the Viet Cong were coming. The helicopter flew me back to Bien Hoa Air Base. It was just a supply ship that had heard my wingman's emergency call. The helicopter pilot watched me bail out and flew down to get me. Otherwise, I would've been captured by the Viet Cong.

John A. Duff,[5] Army helicopter pilot, Vietnam

Everybody was trained and the aircraft were in top-notch shape for the Tet Offensive of 1968, when the North Vietnamese decided to try an all-out assault against the Americans and everybody else in an attempt to turn the

---

[5] John A. Duff was born 16 December 1936 in Paul, Idaho. Following his graduation from high school, Duff attended Idaho State University, where he received a bachelor's degree in architecture. Duff was married and had two boys (ages six and four) and a daughter (three months) when he began his first tour of duty in Vietnam. He had two tours: the first in 1965-66 and the second in 1968-69. Since his retirement from the Army as a colonel in 1984, Duff has been a financial planner in Fairfax, Virginia.

war around. Well, as it so happened, we were loaded, cocked, and ready for them.

They made the mistake of attacking in the north the day before they started in the south, so we had a full day's notification that they were coming. We caught them, and we caught them dead. We caught entire battalions and regiment-sized units out in the open. We were firing not just mini-guns, which were the six-barrelled machine guns; we were also firing twenty-millimeter cannons and rockets. We were using the new rockets that had the fifteen-pound warhead, which was equal in power to a 105-millimeter howitzer, and we were also firing what we called "nails," which were fleshets.[6] As the round came out, it had a proximity fuse and a preset distance above the ground, like about fifteen feet or so, at which the warhead would explode, and it would rain little sharp nails with arrow-type pins into the target area. Unprotected troops were just murdered with those things—literally nailed to the ground. We caught troops out in the open with that kind of ordnance, and it would just slaughter them. So after one day, we took the body count of the troops we'd killed with just three Cobra helicopters. We had a body count of 780 North Vietnamese soldiers in one field. And that was just one day.

It started that day and it ran for about a week, fighting day and night. As fast as we could re-arm, refuel, and get some food and sleep, we were back in the air and shooting people. Finally, at the end of the week they gave out. They ran out of people and ammunition. They ran out of everything, and we chased them for another month-and-a-half back across the border into Cambodia. We decimated them. The Mekong Delta, after that, was practically pacified. There wasn't over a half a dozen incidents per week in the entire delta. So it was a victory for us; yet in the news-

---

[6] Ammunition that splattered shrapnel on impact.

papers it was called a victory for the North Vietnamese, which I could never understand.

David I. Folkman, Jr., Air Force fighter pilot, Vietnam

It was very challenging to get the enemy without getting the friendly troops. During one mission in South Vietnam, two F-100s attacked a rocky hill with caves in it. The Viet Cong were in these caves, and the friendly troops were down below trying to get these Viet Cong out of there. The Viet Cong could see the F-100s coming in and they were firing rockets at them. On the first pass, one of the F-100 pilots dropped his bomb. Afterward you could hear screaming—he'd hit the friendlies. He'd been so afraid to get near the enemy that he'd dropped his bombs short and hit our own people.

J. Keith Melville, Army Air Forces bomber pilot, World War II

The Distinguished Flying Cross (DFC) is supposedly a medal given for merit. There's a certain phoniness to these military awards. Let me explain how I received the Distinguished Flying Cross.

As pilots approached the end of their tour of duty, the public relations officer would say: "Write up what you consider to be your most meritorious mission, the one that displayed the greatest amount of bravery, getting the plane home with engines shot out, resisting enemy fighters, or some other noteworthy combat achievement. We'll then take this and write it up as justification for receiving the DFC." The descriptions of most DFCs had this connotation—"flying home on a wing and a prayer" stuff.

I told my squadron public relations officer that the only mission I'd accept a DFC for was the mission over Vienna when the group commanding officer wanted to court-martial me. The CO with other top brass of the 463d Bomb Group were flying in the lead plane, and our group was leading the entire Fifteenth Air Force over the target. I was the lead pilot of the 774th Bomb Squadron, which was flying off the right side of the lead squadron in typical formation of four squadrons in a diamond-shaped group. Apparently there was confusion in the lead plane as we approached the "initial point" where we were to start the bomb run. They'd let their air speed drop down as they turned to the right to start the bomb run, which meant my group had to cut our air speed even lower to stay in formation. The pilots in my squadron broke radio silence and complained they were about to stall out. I had to make a decision. I could pull out of formation and make a 360-degree turn, but that would put my squadron so far behind my group we'd never catch up before our group went over the target. We also might get in the way of the group behind us. I quickly asked my navigator if he had located the "initial point," or IP, on the ground where the planes started the bomb run, and he said yes. I then asked the bombardier if he could set the target into his bomb sight. When he answered yes, I simply turned onto the bomb run ahead of the CO's squadron and led our group and the whole Fifteenth Air Force over the target.

When the group had its debriefing after the mission, the CO was furious. He threatened disciplinary action for me, but my squadron commander stood up for me and the issue was dropped. I'd made the top brass of the group look bad and possibly denied their chance for glory—and justification for medals. But I was just doing what I had to do to complete a successful mission.

Lawrence H. Johnson, Army Air Forces bomber pilot, World War II

After one of our lone-wolf night strikes against Clark Air Field in the Philippines, we headed our B-24 bomber back to our base in New Guinea. It was only a few miles away from a Japanese air base, so we maintained radio silence. The navigator brought us over the coastline and I spotted a landing field that was very similar to ours, but it just didn't seem right. After flying for several thousand miles in one direction and then going back, it's very easy to be off a couple of miles. So I flashed the landing lights of our bomber, which was the signal to land, and we got a green light from the tower, but it still didn't seem right. So I turned away from the runway and had the navigator and the radar man recheck. It just didn't pan out. Unless my compass was off, there had been a change in direction of the runway. So, we went down the coast and found our base. We'd almost landed on a Japanese runway. On one of our first combat missions, we almost ended up as prisoners.

On another occasion when we were leading a flight of three aircraft in a group formation, the squadron had just left the target located on north Formosa. There was a fair amount of anti-aircraft fire over the target; in fact, one burst put a hole in our B-24. The number four engine stopped about the same time. It was very close to us and made a terrible noise — the copilot jumped about a foot out of his seat. Of course, we still had three engines, but the big problem was we couldn't keep our formation speed. So my flight began to fall behind the squadron formation. Instructions said we were never to stay with a "wounded duck" because that meant you jeopardized other aircraft. If we were going to get picked off as a straggler, don't take others down too.

Despite this, my wingmen stayed right with me. Finally I told them that I couldn't stay up with the formation and they were to go join the rest of the formation. I thought that was the end of our crew. All this time the copilot kept insisting that I try to start the dead engine. That seemed a really stupid thing to do, because you don't try to start a torn-up engine that might blow up the airplane. He never explained why he kept insisting that I try to start the engine. He finally said, with a plea in his voice that rang with sincerity, "I'd try to start that engine if I were you, Johnson."

So I relented, "All right, I'll just try to start it once." We were far behind the other bombers, and there were fighters in the area. It started right up — what a wonderful, amazing feeling. The squadron reduced speed, allowing us to pull up into formation. My wingmen formed on us again.

After we landed at our Philippine air base, I asked my copilot, "How did you know to start that engine?"

And he said, "Well, I might as well admit that when I jumped up in the air, I came down and kicked off the magneto ignition switch with my foot, and that's why it went off."

That really made me angry, because such a mistake could've killed the whole crew. To me, who'd always been told to be forthright, it seemed that he should've grabbed me and said, "Hey, I did that. Let's just start it right up." It's somewhat humorous to recall, but it was deadly serious then.

Richard A. Baldwin,[7] Air Force fighter pilot, Korea and Vietnam

In Korea one time we were dropping napalm, which is a jelly that explodes when ignited with certain fuses and charges. I remember sometimes the canisters would sit out in the rain and water would get down in the jelly and some of the fuses wouldn't fire off when we dropped them. On a particular day I was following one of our aircraft on another course to protect him, but when he dropped his napalm, it didn't explode. So I hopped down very low trying to do a polo mallet-type thing. I followed the napalm and kept it

---

[7]Richard A. Baldwin was born 21 February 1923 in Salt Lake City, Utah. Baldwin graduated from East (Salt Lake) High School, studied political science for three years at the University of Utah, married, and had a son (1948) before going to Korea in 1950. He had a second tour of duty in Korea in 1953. Prior to his 1970-71 tour in Vietnam, Baldwin earned a bachelor's degree in international relations from Syracuse University in 1959, a master's degree in public administration from American University in 1968, and a master's degree in educational administration from the University of California at Los Angeles in 1969. He was a professor of aerospace studies at Brigham Young University from 1971 to 1973. Baldwin retired as a colonel from the Air Force to pursue business development interests.

in sight. You were only about fourteen feet off the ground when you'd release them. I got right on the ground and could see them, so I pushed my nose over and put a .50-caliber machine gun through both of them. I was very lucky, more luck than skill I'm sure, that they exploded. But they exploded right in my canopy, right in front of my nose.

Of course, the great fireball that resulted consumed all of the oxygen that my airplane needed to keep combustion in the back. Going through it, I flamed out—I lost the fire in my engine. I was only fourteen feet or so from the ground and pulling up by this time. I had no life in the thing.

I just did all of the things naturally that you'd do through years and years of combat training. I didn't even think about them. Hands went to switches, throttles came back, emergency fuels went on, emergency ignitions went on, back came the throttle, up went the throttle, and all of a sudden I got a start, which no one had ever done before like that. I got this start, and it almost shook all of the buckets off of the compressor, but I was able to get back home.

Werner Glen Weeks,[8] Army helicopter pilot, Vietnam

Flying during a combat assault mission in the Ashau Valley, my ship was second in a daisy-chain of five helicopters that were taking combat troops into a very tight landing zone, or LZ. The LZ had space for only one aircraft and was in the middle of triple canopy jungle that had been blown away by artillery and bombs sufficient to make a "hover hole" into which we could descend.

---

[8] Werner Glen Weeks was born 8 September 1945 in Woodland, California. After graduating from Provo (Utah) High School and studying business at Brigham Young University for three years, Weeks joined the Army. He was twenty-four and single when he went to Vietnam. Weeks retired as a major from the Utah National Guard in 1989. He is a home preparedness and financial consultant in Salt Lake City.

As we strung our formation out to allow time for each aircraft to descend, drop its load, and climb back out, the lead aircraft reported it had a major malfunction. It had to land as quickly as possible. As the second aircraft of the formation, we now became lead—the first ones to test the flight conditions of the LZ.

Flying under combat conditions always required us to operate our helicopters close to the upper limits of their performance capabilities. Our weights were always near maximum, and we were constantly required to monitor atmospheric conditions to ensure there was sufficient lift to keep us in the air.

As we approached our "hover hole" in the trees, I monitored the gauges while the other pilot flew. We seemed to have enough power to make the two hundred-foot vertical descent to drop off our troops. As we descended below the tree tops we lost that bit of lift the wind had provided, and at maximum power we found ourselves short of sufficient power to remain airborne. In short, we began to fall out of the sky as the RPM began to bleed on our engine and main rotor system. The effect can best be described as similar to the experience of going up a hill in a car in third gear and feeling the power wane until a shift to second is required to clear the crest of the hill. In this case, the lower our rotor RPM became, the more our blades flexed upward and the smaller the rotor disk became. Effective lift was decreasing until we reached a point that a crash was inevitable.

Our operating RPM normally would be 6,600. As we fell through 5,600, I knew we were incapable of sustaining flight. I left the gauges inside the aircraft and looked outside to find a place to crash "softly" in hopes of preserving all lives on board.

At that moment, when I knew death was possible, suddenly a voice came to me. I could hear it very audibly say, "Glen, let your heart be at ease, and watch." My terrified heart was suddenly calm. My spirit became very meek.

With both feet on the floor of the aircraft and releasing my grip on the seat, I folded my arms and watched.

There was nothing I could do to aid the other pilot. His hands were full trying to maintain aircraft control. Then, not only did we stabilize, a miracle in itself, but we climbed. Coming up out of the hole and on our way, I knew that he understood the extent of the divine help we'd received to get out.

We relayed by radio the conditions we'd encountered, warning the other aircraft of the dangers. But the seasoned veterans who followed us went into the same landing zone, and all suffered damage for their decisions. That day my aircraft was the only one that came back undamaged.

Grant Warren,[9] Air Force rescue, Vietnam

There was a naval lieutenant commander stationed on a carrier out in the gulf off Saigon. We went in and pulled him out. This guy was as fast as you've ever seen. He must have set a record for the fifty-yard dash when he saw us. We picked him up and hauled him up right out of the clutches of the VC. He was a crazy guy. He was jumping for joy and hugging us.

We flew to the base and dropped him off. We called the carrier and a navy chopper came and got him. About three or four days later, here was this naval chopper sitting

---

[9] Grant L. Warren was born 12 September 1947 in Southgate, California. Before Warren's service in the U.S. Air Force began at age twenty, he graduated from Fairview High School in Boulder, Colorado, joined the LDS church in 1964, and earned an associate's degree from Chaffey Junior College in Alta Loma, California. After his Vietnam experience, Warren worked for some time as an executive for the Boy Scouts of America. He currently owns and operates a Lacey, Washington, business that provides community-based residential services to developmentally disabled adults.

down on the runway. Our commanding officer said to me and the other guy I flew with, "That's for you guys. You get out there and get on that."

They flew us out to the carrier. The guy we rescued was waiting for us there with champagne. He said, "Come out here, guys."

It was embarrassing being LDS. I said, "Well, sir, I don't drink."

He said, "What do you want? Coke, Pepsi, 7-Up, water, anything! I've got it." He had this ice chest on his helicopter with all of this stuff in it. "You guys are great," he said. We had T-bone steaks and baked potatoes and saw a movie and sat in air conditioning, which was the greatest. That guy was eternally grateful to us.

Dennis E. Holden, Marine infantryman, Vietnam

We were flying back by helicopter to main camp. We passed over the Arizona territory, which was Indian country—it was where all the bad guys were. We were hit with machine-gun fire. That put the helicopter into an auto rotate spin, which means it was going down to crash. That was probably the most vivid thirty seconds of my life. A lot of panic goes on in front of your eyes. I remember I was sitting next to a Catholic chaplain who had his rosary out during all the panic and screaming. It was like a slow-motion movie. The priest was praying and crossing himself. I reached over and grabbed the rosary and tried to cross myself, and I wasn't even Catholic.

The helicopter was burning when we crashed into the river and it immediately sank. As far as I could tell, I was the only survivor. I was able to make my way to the nearest bank that had a grassy over-hang. I knew the VC would be looking for our helicopter and any survivors. I lay in the water, completely concealed under this bank, and waited for the VC to arrive. Within an hour, the entire area was being combed by VC looking for survivors. The second day they began salvaging the helicopter ammunition and supplies from the wreckage while they continued to search for survivors. I didn't move for two days out of fear for my life and fear of being captured. On the second night the VC left, and I climbed out of the river and started working my way south, hoping to run into friendly troops. I spent the next two days hiding in the jungle during daylight and moving slowly at night. On the fourth night, an area to the west was bombarded with mortar fire. My instincts told me this was enemy fire and that there must be friendly troops in the area. At first light, I moved to the shelled area and, to my relief, found a patrol from the Second Infantry.

Ted L. Weaver, Army Air Forces bomber pilot, World War II

When I first got there, 25 percent of our aircraft were lost every mission. This meant that after I flew four missions, as long as I kept flying, someone had to go down in my place. This made the odds seem a little bit tough. We were told by the army that if we dropped two successful bomb loads, we'd paid for our training and our ship. Therefore, anything we dropped after that was on the black side of the ledger. It wasn't very encouraging to feel like you were that expendable.

At the time, we were required to fly twenty-five missions to fulfill a tour of duty. If 25 percent were being shot down each mission and you fly twenty-five missions, you don't have a very good chance of getting through.

Richard A. Baldwin, Air Force fighter pilot, Korea and Vietnam

We had what we called purple routes. They were the most deadly of all of the combat assignments. We were on a purple route, and I was the number four man of a crew of four because I was new. The flight leader and his number two man—his "element"—were flying at about two hundred to five hundred feet, whatever altitude it took for them to see what they could. They were cleared to use rockets, bombs, napalm, and .50-caliber machine guns.

They were down there for a little while, and then

all of a sudden I felt the strangest thing come over me. I had this premonition. I looked up over my shoulder and I saw twenty-four planes bearing down on our group, but they hadn't seen me. I did what any fighter pilot would do. I yelled into the microphone as best I could, "MIGs at three o'clock." They were coming right down on my flight leader and his element. "Break, Break!" I yelled.

At the same time they did this, I broke right into the center of their formation, firing my guns, forgetting at that moment I needed to try to get rid of my tip tanks. I was firing and flying, trying to dip my tip tanks, I couldn't get them to drop. There I was, sitting with four hundred gallons of fuel on the end of my wings, which were absolutely an albatross around my neck. For some reason or another the device that blew them off wasn't working.

So I took the aircraft and just put torque one way and the other, big heavy wing drops over and under, and literally threw them like a slingshot as far as I could until they finally came off and I registered the maximum G-forces. I looked out, expecting to see my wings all crooked and warped where these tanks came out. But they were okay, and I was able, then, to fly. But this caused me to lose my leader, my element leader, who was the number three man in the flight. I told him, "I've lost you. I've got my tip tanks off now. Have you?"

He said, "Yes."

I said, "I'll try to join up with you where I can."

I just took off after the MIGs. The one thing about the MIGs was that we could turn inside of them in a slower airplane, which gave us the advantage. We went at them and fought them. I got very lucky and got a couple of them. That night it came over the national news that our flight had been shot down by the MIGs.

LaVell Meldrum Bigelow,[10] Navy fighter pilot, World War II

One of the initial strikes of the battle of the Coral Sea came after we sighted the Japanese carriers as they were coming down from the north. This carrier force was well-protected by fighters and anti-aircraft fire. Our group of eighteen dive bombers approached the Japanese ships at an attack altitude of fourteen thousand feet. Just as I was nosing over into my dive, a Japanese Zero came up on my tail. This was going to be a very hairy attack with this Zero on my tail and with the anti-aircraft fire. I had to make an immediate decision whether to continue down in a steady dive or to take evasive action, which would ruin my aim and probably force me to miss my target. So while I was in this dive I thought, "I'll continue my dive and rely on the Lord to protect me from being injured." So I persisted in my dive and held it steady. As far as I know, that Japanese pilot didn't fire one shot. If he did, I didn't see it, and nothing hit my plane. He followed me all the way down through the dive without shooting, and my plane wasn't hit with anti-aircraft fire either. My bomb struck the Japanese carrier. After I pulled out of the dive, he went his way and I went my way. For that particular attack I was awarded the Navy Cross. But it was also a lesson to me that I needed to rely on the Lord and put my fate in his hands. I went through the entire war without being injured or having my plane hit.

---

[10] LaVell Meldrum Bigelow was born 12 November 1917 in Provo, Utah. After attending Brigham Young High School, graduating with a bachelor's degree in physics and geology from Brigham Young University, and marrying, Bigelow joined the U.S. Navy following the United States' entrance into World War II. After his combat experience he earned a doctorate in business administration. Bigelow retired as a captain from the U.S. Navy and now works as a business consultant, a part-time university teacher, and an organ builder.

John A. Duff, Army helicopter pilot, Vietnam

I was shot down a total of four times. Two of the times I was shot down because I made dumb mistakes. The other two times I was shot down just because I got trapped — no fault of my own, but because we were outgunned.

The first time I was shot down was one of those late afternoons when the VC started to shoot at us. Out my right-hand door I could see where "Charlie" was. I could see the muzzle of his gun flashing. I kicked in the right rudder pedal and took it into a very tight turn to be able to bring my guns around to bear and let the door gunners and crew chief fire at the VC. But in so doing, I lost all forward momentum and made myself almost a stationary target in the air. So even those poor slobs down there who didn't know how to shoot at helicopters could shoot at us, and they poked several great big holes in the aircraft.

Of course, all the caution panels came on, the engine quit, all the beepers started flying, everybody started screaming. We were leaking fuel and hydraulic fluid, and we were about 150 feet in the air. All I hoped was that I could just get it on the ground in one piece, hopefully without getting everybody killed. So I straightened it out and brought it over. I knew where the bad guys were, and I didn't want to land next to them. But I didn't have too much of a choice, because we were practically on top of them when they hit us. So I just put it down out in the middle of a rice paddy, which happened to be right between the good guys and the bad guys; right in the middle of the firefight. As we touched the ground, we jumped out into the rice paddy, which was about two feet of water and about a foot-and-a-half of mud. Fortunately, no one was wounded. We were up to our necks and hiding behind rice stacks. Then we just slithered on away from the ship back over toward the good guys.

When we were hit, they brought in the Air Force and began air strikes, and it got dark, and "Charlie" went away. The big Chinook helicopters came in and lifted out the broken aircraft and us and took us home. Fortunately no one was wounded. As for the aircraft, the only damage was where a bullet had penetrated the transmission. The next day, it was up and flying.

The second time I was shot down, we were flying in support of a ground operation due west of Saigon in an area notorious for having people in concrete bunkers who would shoot at anybody. So they sent in the ARVN [Army of the Republic of Vietnam] ground troops to try to root them out of there, and I had my fire team flying fire support for it. But the enemy was shooting back, and I should've been smart enough to see that our suppressing fire wasn't doing any good. The big orange balls were coming up faster than ours were going down, and one of them hit the ship. It killed the door gunner, wounded the crew chief, blasted me with shrapnel in the arm and side, and blew plexiglass all through our faces. Fortunately the copilot wasn't injured and the airplane was still flying. So I gave control to the copilot, and we broke off and got back to the aid station.

The third time I got shot down was when we had to patrol the water one night. We had .50-caliber or 12.5-millimeter shells hit the gun ships we were flying. This blew them all to the devil—blew the front canopy off, hit the transmission, put two 12.5s through my instrument panel, blew all my instruments away. The gunner up front was wounded, and the bird was pretty well worthless. All we could do was just set it down in the jungle in bad guy territory. As soon as we hit the ground, we were under fire constantly from all sides. I pulled the gunner out and brought him down to the dry paddy. I applied the tourniquets as best as I could with belts and a first aid kit. Then we waded through the rice paddy over to the canal.

The bad guys searched the canal, but somebody was watching over us, so they didn't find us. We stayed there until the next day. With the battle still continuing, I crawled up a side-running ditch and got the gunner out of the water. I could see that he was dead, so I put him underneath some bushes. Finally a good guy helicopter arrived and pulled us out.

The fourth time I was shot down, I was flying cover for a small operation up along the border of Cambodia. I didn't realize the bad guys were there. We were flying very close cover for "families" as they were walking through the tall elephant grass, and I was keeping visual contact with them because we didn't want to shoot the wrong people. What I thought were good guys turned out to be bad guys, and they caught me in a crossfire. They had a habit of setting up three .50-caliber machine guns in a triangle as a helicopter trap, and sure enough, they trapped me in a crossfire between those three guns. They literally blew us out of the sky. The tail rotor came off. We crashed to the ground and the ship rolled over on its side, tearing the lead rotor blade off and breaking the canopy. We moved back away from the bad guys and hid in the elephant grass until the good guys got to us and took us out.

Ray T. Matheny, Army Air Forces flight engineer and gunner, World War II

Within about five miles of the target area I saw a B-17 on our left coming opposite our path. Harper called out the B-17 on the intercom as a "friendly" passing by. I thought it was mighty peculiar that a B-17 crew would fly in that direction. I adjusted the reticles on my gun sight to seventy-five feet for a fuselage length. I waited. Then suddenly bright flashes appeared at the port gun positions. "I knew it," I exclaimed, "he's a Nazi!" I brought the gun sight down and framed the fuselage. I could now see an Iron Cross painted on the fuselage where the American star insignia was supposed to be. I fired a long burst as he passed but quit firing because in tracking him I risked hitting other B-17s that were nearby. Harper let out an oath over the intercom and said it was one of our ships. I yelled back that it had Iron Crosses on it. Dunning said that my guns had broken his windscreen because I'd fired when my guns were angled over this head. I was disappointed when nobody else fired on the enemy B-17, and I still had a running argument with Harper.

Richard A. Baldwin, Air Force fighter pilot, Korea and Vietnam

I had a good young friend when I first arrived in Japan. I was on my seventh mission and he'd already finished his tour of duty. He was a second lieutenant, a bright-eyed fellow, as I remember him. I was living in the same tent with him. He had his bags all packed and his orders to go home. He was just sitting waiting to leave the next day. A call came down for a pilot. They had someone drop out and they needed a spare. Someone couldn't fly for some reason.

So this young man, against orders, jumps into a flying suit, runs up to operations, gets in the airplane, and flies up to Taegu, which was where we operated out of for our fuel and our rearming. Then he flew all the way up to the Yalu. There were MIGs there that day and he caught a MIG between a cloud level of about thirteen hundred feet. This MIG was just cruising along, not doing anything. He didn't think anybody would be up there flying. He was just getting a little good weather time before he went back down to the Yalu and landed.

This young fellow was in a flight of four airplanes. We had a problem with our guns. If we didn't get the gun heaters on soon enough or if they weren't working right when we pushed the button to fire them, the bullet just kind of went out the end of the gun. We couldn't shoot them. It was really a bad situation for us. It was cold up there, so we had to make sure our gun heaters were on before we ever left the ground so that they'd be warm by the time we got up.

So here's the flight leader of that airplane getting behind this MIG and trying to squeeze him off. He said, "I can't shoot. My gun heaters are acting up."

This young lad said, "Move out of the way and I'll come in and I'll get the MIG."

The MIG hadn't seen them yet, and the number one leader didn't want to lose the chance to shoot down that MIG. He said, "No, give me a little while, give me a little while." They were arguing in the air. We could hear them.

Finally the leader laid out. He couldn't play it much longer because they were getting too close to that MIG. This young fellow said, "I'm coming in, get out of the way." The MIG was up in front and the flight leader was just closing in on him. He pulled way inside the flight leader, and of course he was now more visible because he was more up front and off to the side of the MIG. The MIG pilot evidently looked over his shoulder and saw this airplane coming. He broke back to the left and had a mid-air collision with this young pilot who had flown his last mission and was on his way home.

He'd written home and told them he was coming and was all packed. Then he flew one more mission. He finally got his MIG, but the MIG got him, too.

C. Grant Ash, Army Air Forces bombardier, World War II

I remember seeing bursts of flack above us. "Hey," I said, "these guys are not very good." Then somebody said, "They aren't, are they? They're even bursting below us now." The next burst went right through the middle of us. They cut us in half.

I thought the boy in the turret was dead. He didn't move. I could see him through two small windows, but I had to open two doors to get to him. He was facing straight forward, so I couldn't see his face. He didn't move. I watched him while I put on my parachute.

But he wasn't hurt. He was just sitting there. The explosion hadn't left anything around him. All he had were the handles on his turret and the freezing cold air at eighteen thousand feet blowing in his face. I remember one .50-caliber machine gun was bent like a pretzel off to the right and the other one off to the left and down. I pulled him back in and closed the doors. I snapped his chute on him and opened the nose-wheel doors. I said to him, "Are you ready to go?"

He said, "Yes."

I said, "Follow me." I bailed out. I turned around and saw him bail out after me.

Ray T. Matheny, Army Air Forces flight engineer and gunner, World War II

The weather was miserable, with a scattered ceiling around twelve hundred feet. It was pitch dark when we took off. Airplanes were trying to find each other and to avoid the low-lying clouds. There seemed to be more confusion than ever, and our crew kept calling out nearby planes to avoid a collision. We'd gotten to about two thousand feet when I saw planes blinking their tail lights in code and using their wing position lights. My adrenaline output sped up as the encircling lights and dim, ghostly outlines of ships in the changing light milled close to each other. Then it happened.

I saw two sets of lights and the faint outlines of two B-17s going in opposite directions. I tapped Dunning on the shoulder and pointed, saying nothing. The two sets of lights then swiftly came together, culminating in a huge fireball, and then the fiery wreckage of both airplanes separated and dropped to earth, where there were more explosions and fires.

"God!" exclaimed Harper.

"Oh, Jesus," gasped Dunning.

I thought of the twenty men who had just died instantly, without a chance to escape the holocaust. We saw other planes veer away from this deadly course. It was an accident we'd feared many times.

Our flight was to take us over the North Sea, and I thought of another B-17 that looped three times before plunging straight down and the men who bailed out over that deadly cold water. I didn't like any part of these flights, but there remained a sense of duty and the righteousness of our involvement.

The Friesian Islands appeared. This coast had become familiar to me these past two months. There was always a feeling of anticipation when we entered airspace over German-controlled land. No enemy yet.

North of Hamburg I got the urge to go to the bathroom. The pains were persistent, but the only thing I could do was grit my teeth. Finally we were over the Baltic Sea going northeast, then we began the slow left turn over Kiel Bay. The urge to go became more painful. I tried concentrating on the unusual condensation trails that floated in the rarified atmosphere for hundreds of miles. I looked at the free-air temperature: it was minus fifty-six degrees centigrade. That was terribly cold and required me to cycle the propellers every fifteen minutes. I checked on everyone's electrically heated underwear to see if anyone needed the shorting wires I had ready. All were okay, but the pains persisted. I began to fantasize about G-8 and his Battle Aces and the stories of Spads[11] against Fokker D-7s[12] in World War I. The pulp magazine of G-8 had been one of my favorites through high school. The pain in my bowels persisted.

Now near the target came the first flak. Our group flew the low position in a combat wing staggered laterally and stacked at about three thousand feet in three groups. Our ship was the last and lowest airplane of the combat box. There were about forty planes in this mighty armada, each streaming lazy contrails of ice crystals. The flak wasn't accurate enough to bring any ships down, but the black puffs were pretty close.

No fighters. Perhaps our diversion tactic had worked after all. The bomb run was made in clear weather and the

---

[11] Spad 13s were French planes used in World War I.
[12] Fokker D-7s were German planes used in World War I.

bombs got away without serious determent. Now we'd head for home.

"One-oh-nines high at six o'clock," came Bigner's voice, followed by a call of more from the east.

"Where's the famous fighter cover?" I wondered as I cranked the fuselage dimensions of the Messerschmitt-109 into the gun sight. High above us, at about thirty-five thousand feet, I judged, were contrails made by planes flying in slow circles. I kept my eye on these distant planes. An Me-109 came from abeam and I was ready, tracking and firing short bursts. He did a split S and dove away, but no smoke. Another came in high astern. I switched the sight to wing dimensions of thirty-two and one-half feet. The fighter's nose showed sparkles of fire from its two thirteen-millimeter machine guns. The pilot would use the tracer bullets to range the target. Then as he drew closer, both wings near the roots lit up — the cannons were firing their deadly load. I could see thin smoke streamers overshoot our right wing. I fired and tracked as quickly as I could. The fighter disappeared but would attack again.

I glanced upward to keep track of the high flyers. They were in close enough for me to see that they were American P-51 and P-38 fighters. I announced the good news over the intercom. I looked at my faithful Piccard watch: it was 11:30 hours. The escort would be with us until we left the coast.

The fight continued. "Why don't the fighters come down here where the enemy is?" I mumbled. Then I noticed Bigner's guns firing astern. The burst was long. I glanced back in time to see his gun barrels come to rest pointed upwards. He'd dropped his guns. Why? My turret was swinging around when I saw the Me-109 bearing on the tail and firing all guns. The Me-109 pilot had shot our tail gunner and was still coming. I was trying to get my sight on him when I saw Smith's radio post gun barrel suddenly go up to the rest position. "Smith's got it too," raced through my mind. "I've got to get that Jerry." I was about to press the triggers when the Me-109 disappeared into our contrail. "So that's his game," I thought. "I'll be ready next time."

My sight was set on the dimensions of the Me-109's wings and I'd cranked in a range of about two hundred yards in anticipation of a renewed attack using our contrail as a screen. We were the lowest ship and had no protection from the others. I called Tex in the ball turret to alert him to the Me-109's tactic. He was busy and didn't answer.

I hadn't long to wait. The nose of the Me-109 came up suddenly, and as the pilot was leveling off, I was adjusting the reticles of the turret sight to fit his wing span. This target was changing speed, and as he drew closer I had to move the bicycle-type controls to properly frame him. I began firing at about the same time he did. I could see the flashes of his guns so close now, maybe seventy-five yards away. I fired a long burst. The right gun quit firing. I kept firing the left gun. I saw smoke streamers from twenty or thirty-millimeter cannon shells passing a few inches over my turret. It was a duel now, both of us firing point blank.

The Me-109 was close when I saw pieces of the engine cowling fly off. The 109's guns quit firing, but the plane kept coming. I saw the propeller of the 109 almost stop as I poured steady fire into it. The 109 was now about thirty feet away and had coursed slightly up and to the right. "I killed the pilot," I thought, "and he's going to crash into us." I yelled into the intercom, "Jump it," for evasive action. The maneuver command was for both pilots to pull back suddenly on the control column, then push down to avoid crashing into planes above and to miss an oncoming aircraft. I anticipated the maneuver by placing my forehead solidly against the gun sight, hunching down, and throwing both arms up over the tops of the ammunition cans to keep the ammunition from being thrown out through the top of the turret when the pilots would push down on the control column.

I wasn't surprised when the explosion knocked me

out of the turret onto the floor. I thought, "God Almighty, he rammed us. The ship's going down." Our ship rolled on its back, then came around right side up. I grabbed my parachute, which I'd attached to the longerons with break-away safety wire. I quickly fastened the right hook of the chest pack onto my parachute harness. Then, to my amazement, the parachute was drawn out of my grasp. I began to be pushed to the floor as G-forces started to build up. We were in a flat spin. I rolled on the floor and tried to get the left parachute hook snapped on. I glanced up and saw Harper trying to reach his parachute, which was hooked to the back of his seat. I saw Dunning vainly trying to control the plane so the others could bail out.

The long oxygen hose I'd installed earlier kept me alive as I struggled on the floor. I rolled again, this time into the crawlway joining the nose section with the cockpit. I crawled forward on my belly through control cables and wires that were drawn out of place by the centrifugal force of the flat spin. Black smoke obscured my vision. With great effort I crawled up to the little escape hatch on the left side of the fuselage next to the number two propeller. It would've been so simple just to lie there and die, but some force within kept me struggling, like a dying animal in a trap. I saw Doty lying on his back, wearing his steel helmet, flak suit, Mae West,[13] and flight gear, with one hand stretched out trying to reach the hatch handle. I couldn't help him. His long-standing premonition that he wouldn't survive combat was being fulfilled.

I grabbed the red handle of the escape hatch and pulled the hinge pins out. The door should've flown off into space, but the twisting of the fuselage in the spin must have held it in place. I had no thoughts of helping Doty; it was a case of putting all my remaining strength into saving my

---

[13] An inflatable, vest-like life preserver.

own life. The last thing I remember was pulling down the hatch latch handle and beating my fist on the door.

My next conscious moment I was free-falling, and my chest pack, still hooked only to the right side, was beating me in the face. I knew that I was falling and free of the airplane. I pulled at the D-ring without regard for how high I was. The twenty-four-foot canopy spilled out with a shock that hurt my chest. As it turned out that day, Nelsen had gotten my chute harness and I'd gotten his. Nelsen was about twenty-five pounds heavier than I and his harness was loose on my body. Also, the single hook on the right side of the harness pulled it sideways, displacing all of my left front ribs when my chute opened. Nevertheless, it was a beautiful sight to see the white silk canopy of the parachute.

My consciousness of what was happening came in stages. First, I noticed that I could no longer hear the roar of the bombers. A few seconds earlier, I remembered, there were forty bombers in the heat of battle, each spitting out an awful noise of engines and propellers, plus the chatter of guns. Now all was silent, and I looked up in vain for the bombers. They were gone! Then I realized that my free fall had been a long way—thousands of feet—and that the bombers and fighters were traveling away at 250 miles per hour while I was going down. But I could still hear my plane making the terrible sound of overspeeding propellers and I saw the flaming wreck gyrating down. I let out a choked, painful cry, "Please, God, let them live." But I knew that it was too late; the crew was dead. That was my first plea to God for anything and I didn't expect a response; I only hoped for one.

I saw a spinning, smoking fighter plane crash in the distance. It looked like an Me-109, but I couldn't be sure. Then it began to rain small pieces of airplane. I could see the pieces hit the silk canopy, then slowly slide off the rounded surface past me. One side of the metal was o.d. [olive drab] war paint and the other was light green, zinc chromate. I became aware of a piece of burning wing drifting away from me as it flipped over and over. It looked like the outer panel of a B-17 wing. In what sequence all of these events took place I don't know. Undoubtedly several were going on at the same time, but my consciousness would only let me focus on one at a time.

I looked below me. I could see about four to six power lines and figured they each carried high voltage. I thought, "How did they teach us back in Texas to guide a parachute? Oh yes, we had fifteen minutes of parachute instruction that day. Let's see, the instructor said something about pulling on the risers."

I pulled on the left riser and the chute responded by swinging me like a clock pendulum. I tried again, but with the same result. The wires were getting closer.

I could see a small village strung along a road made up of eight to ten houses. A few people were moving toward where I figured to strike the high-tension wires. There was another road below me covered with snow. The wires came up fast, and in desperation I pulled down on the left riser until the chute collapsed. The parachute let me straight down, past the wires. I let go of the risers and the parachute popped open just in time to break my fall. I landed in a canal and broke through the ice. The canal apparently had been nearly drained for the winter and had only about two feet of water in it. I lay there on the ice with my legs submerged in the water, greatly relieved but with a sense of burden—a burden I didn't understand. This ended my short career as a combat flyer in the Eighth Air Force.

Grant Warren, Air Force rescue, Vietnam

I really hadn't given much thought about fear or anything until I actually got there and went on my first mission. I went with a brand new pilot, a first lieutenant Air Force pilot, and two other fellows.

We went out and we picked up this Navy pilot. We'd identified where he was. I saw him running to the bushes as he was being chased by several Viet Cong. We lowered the hoist collar. He came running and dove at the hoist collar. We immediately started the hoist going up. As we were bringing him up, he was shot several times. He took about eight rounds of AK-47.

When we got him in the chopper I immediately started intravenous fluids. By the time we got back to the base he was dead. I was the medic in charge on the chopper, and the pilot and I had to get the dead pilot's identification so we could officially transfer the body to this army doctor, who would send him to the morgue. I reached into this guy's pocket and took his wallet out. As I took his wallet out, a fold-out picture album dropped out. I can still remember the picture of him standing there with his really nice-looking blond wife and his blond little girls. It blew me away. That was my first mission. I immediately got all shook up and vomited all over the flight line. The lieutenant also vomited all over the flight line. I was a non-drinker, but I'm sure that the lieutenant went out and got thoroughly smashed to forget it. I didn't have anything to do, so I went back and just tried to forget it and tried to talk to some people about it.

PHOTOS
★ ★ ★

"Two of my cousins came back in caskets. . . . Another cousin came back shell-shocked." Pictured here with several friends in France, Ivan A. Farnworth (third from left) died in 1989 at 92 years of age.

"I told my squadron public relations officer that the only mission I'd accept a Distinguished Flying Cross for was ... when the group commanding officer wanted to court-martial me." J. Keith Melville (front row, first on left) with his crew.

"I just participated and wanted to do my part to make sure that the things we treasured and enjoyed as Americans would be preserved."
Clyde Everett Weeks, Jr.

"The navigator brought us over the coastline and I spotted a landing field that was very similar to ours, but it just didn't seem right." Lawrence H. Johnson

"I could never feel any hatred towards anybody, French, British, or American." Walter H. Speidel

"Our sergeant was also untouched because he knew how to delegate authority. [He] never left the command post." David L. Evans

"They brought in some Red Cross stationery and said we could write home.... I just put in big bold letters, 'Who are you monkeys trying to kid?' Ten minutes later . . . I took another working over". G. Easton Brown (front row, second from left) with his crew.

"Just as I turned my head to the right, I got the bullet." Jay Dell Butler, at a military training camp near Little Rock, Arkansas.

"I knew that adultery would be one of the biggest temptations I would be confronted with while I was in the Navy. . . . She expected a lot more than what she got." W. Howard Riley (center) with two friends.

"We knew what we were fighting for! . . . We were fighting for freedom". C. Grant Ash (front row, third from left) with his crew.

| 1 | 2 | 3 | 4 | 5 | 6 | 7 | 8 | 9 | 10 | 11 | 12 | 13 | 14 | 15 | 16 | 17 | 18 | 19 | 20 | 21 | 22 | 23 | 24 | 25 |

Personalkarte I: Personelle Angaben

Kriegsgefang. Lager Nr. 3 d. Lw. (Oflag Luft 3)

Beschriftung der Erkennungsmarke
Nr. 5258
Lager Kgsgef.lo.d.Lw.3

Name: ASH
Vorname: Cecil G.
Geburtstag und -ort: 27.11.22
Religion: Mormone (L.D.S.)
Vorname des Vaters:
Familienname der Mutter:

Staatsangehörigkeit: U.S.A.
Dienstgrad: F/0 1
Truppenteil: USAAF Kom. usw.:
Zivilberuf: Berufs-Gr.:
Matrikel Nr. (Stammrolle des Heimatstaates): 7123697
Gefangennahme (Ort und Datum): Österreich
Ob gesund, krank, verwundet eingeliefert:

Nähere Personalbeschreibung

Grösse	Haarfarbe
1.77	melond

Besondere Kennzeichen:

Fingerabdruck des rechten Zeigefingers

Name und Anschrift der zu benachrichtigenden Person in der Heimat des Kriegsgefangenen

450 North 1st West, Lehi Utah

Lichtbild

Des Kriegsgefangenen

ASH, C.G.
5258

"The first thing the Germans did was put me in a small single room in solitary confinement". C. Grant Ash's prisoner of war information card for his internment in Luft Stalag 3 near Sagan, Germany.

"The camp was such that they allowed you a certain freedom inside restricted compounds." Prisoners play a hockey game at Luft Stalag 3 where C. Grant Ash was held until a forced march began in January 1945.

"We saw one glider . . . that was on fire before it ever hit the ground. When it hit the ground it just disintegrated." Dallis A. Christensen in Germany.

"Three days later the base got an irate call from one of the local English ladies who had gone out to hang her clothes and had found this bomb between the house and the clothesline." Ted L. Weaver

"Our outfit overran a Nazi concentration camp. . . . The Nazis had lined them up and shot them, and the bodies were still laying there." Freeman J. Byington in front of some bombed-out barracks that had housed German SS troops, describing the scene he photographed in the next picture.

"My squad got right up next to the wall. We took a bayonet and put it on the end of a gun and poked a helmet up over that wall, and it got blown off immediately." Lincoln R. Whitaker (on left) with a friend in Germany.

"I was awakened with a start. 'The jeep, the damn jeep is coming' went through my mind and I hoped it was a dream, . . . I dug my fingernails into my palms and silently cursed. Yes, another mission call at 0200 hours." Ray T. Matheny

"I remember the uncertainty of the natives. They had little paper flags and they weren't quite sure ... whether to put out a flag for North Korea, for South Korea, for the United States, for the United Nations, or ... for China." David R. Lyon (fifth from left) with a medical team from India.

"He had his bags all packed and his orders to go home.... Then he flew one more mission."
Richard A. Baldwin

"I can remember I got home late one night from covering the primary elections for KSL and there was a note from Uncle Sam in my mailbox. It was a great shock." Lynn Packer in Vietnam.

"I looked out the right side. Half of my wing was gone, cut right in half. My gas was flowing out and I knew there was no way I was going to make it home." David I. Folkman

"That's where I got wounded. I suppose I'd call it a premonition that I ignored." Douglas T. Hall (back row, third from left) with part of his American-Montagnard Special Forces team.

"We'd just talked to some prisoners we'd captured. I thought, 'What a waste.'" J. Tom Kallunki, describing the experience of taking these Viet Cong prisoners.

"Every day was true adventure. It was an adventure in life and death.... I didn't have time to be scared. I was too busy ... keeping other people alive." Pat Watkins (on left).

"I knew my wife was a strong and capable woman and could manage in my absence. But . . . it wasn't easy to leave a single parent with a house full of teenagers and younger ones to care for." E. Leroy Gunnell

"There's no glory in war, not a bit." J. Tom Kallunki

"Frankly, I didn't see any freedom. All I saw was poverty, waste, sickness, prostitution, drug soliciting, and all of that." Grant Warren (Photo by Danny Foote)

"It's crunch time. [We] need to make a decision. We've got an armada on its way in and its about thirty seconds out from landing. Are we going to land or not? I look at him and he looks at me." Don G. Andrews

"A friend of mine, Captain Monte Lorrigan, and I spent many hours together. He was . . . president of the Sunday school in our branch on Tan Son Nhut Air Force base." David L. Gardner (on right) with Lorrigan.

"The fourth time I was shot down, I was flying cover for a small operation up along the border of Cambodia." John A. Duff

"I really felt, and still feel, that communism is enough of a threat to humanity as a whole that it's worth the loss of life to oppose it. Freedom is worth fighting for." Richard P. Beard

"There was a lot of frustration in [Vietnam]. . . . One just wondered why we couldn't use our power to do what we'd been trained to do and eliminate the limiting, self-imposed constraints on how we fought." Kirk T. Waldron (on left) and his crew — transporting Bob Hope on a USO tour.

"I can remember a Christmas where I just had to go hide because everybody was either stoned or drunk, and if we got overrun, everybody would be killed, so I went and dug a hole and hid the whole night. It was a lonely experience." Danny L. Foote

"It was a badly wounded marine, a radioman, calling for help. The battalion commander personally came up on the radio, . . . 'Now son, hold on; we're coming to get you. Where are you?' A faint reply, 'I don't know. I'm all alone.'" Howard A. Christy (on left)

"They blindfolded me . . . , but I could see just a little bit. . . . They were trying to get me to stand on the [American] flag, so they could get a movie picture of it. Just as I got to the flag, I collapsed and leaned over and kissed the flag." Jay R. Jensen

"We brought in all of our security and we were walking down a road just like farmers going to work. We walked right into an ambush." Hyde L. Taylor (on right)

"When we were walking through My Lai, there were a few villagers running here and there. A lot of them were still trying to get out of the area." Mike Terry at bat.

"As far as bonds, . . . you never really knew any of these guys for longer than six months to a year, so although you call yourselves friends, close friends, it isn't like somebody you grew up with or went to school with." Danny L. Foote (second from left)

Wounded men on a military transport in Vietnam. (Photo by Kirk T. Waldron)

"On 9 July 1972, I had my singularly most anxious experience."
John C. Norton, Jr.

four
KILLING AND BEING KILLED

Jerry L. Jensen, Army Special Forces, Korea and Vietnam

I can remember the first time that I was in a combat situation. The Chinese made a big sweep at us in a big mass attack. I was in my foxhole up on a ridge line. Korea is a mass of ridges. We were dug in just 150 meters below the crest. They came swarming up at us. We were told to hold fire until we received the command. They just kept getting closer and closer. I was watching the eyes of this one guy that was heading right towards me. Finally they gave us the command to fire. He and I fired simultaneously and he missed. I got him. That was the first time I killed anyone.

Robert M. Detweiler,[1] Air Force pilot, Vietnam

We used to drop fire weapons on enemy supply trucks, and we knew through our intelligence that those

---

[1] Robert M. Detweiler was born 20 July 1930 in Centralia, Illinois. He attended Zeigler Community High School in Illinois before receiving an appointment to the U.S. Naval Academy in Annapolis, Maryland, where he earned a bachelor's degree in electrical engineering. Detweiler was thirty-seven years old, married, and had two daughters and a son when he entered combat in

drivers were shackled to the steering wheels in those trucks and couldn't get out. Once those trucks caught on fire, the drivers just burned to death in the trucks. That didn't stop me one bit from going after those trucks and getting those guys. I remember once stopping a convoy of about forty-seven trucks, and we got every truck in that whole convoy. And the things exploded, exploded, exploded all night long. What we got was a convoy carrying weapons and ammunition down south. By stopping the traffic and the flow of supplies to the south we protected all of our troops that were in the south. Just think of all the lives we saved. We killed the enemy, but the guys we saved were our own guys, and that's the whole idea.

George E. Morse,[2] Army Special Forces, Vietnam

I've seen young men who learned to love to kill and maim. They were animals. I couldn't believe some of the things that I saw. I'd think, "This can't be. They must not have been raised like I was raised." I don't know whether it was the war or something else, but some of those guys just went wacko. Maybe it was because of training. The government trained us to be professional assassins and that's what we did.

Vietnam. A former Lutheran, Detweiler joined the LDS church in 1976. Using his master's degree in nuclear physics, he has worked for several organizations as a scientist and research administrator.

[2] George E. Morse was born 4 October 1945 in Torrence, California. After graduating from Provo (Utah) High School, Morse became a member of the U.S. Army Special Forces as a single nineteen-year-old. He had three tours of duty in Vietnam: 1966, 1967, and 1969. Morse is a power supply manager with Provo City.

Peter Bell, Army Special Forces, Vietnam

I saw guys in combat in my unit who delighted in taking life. Often I found myself caught up in the thrill, just like deer hunting, chasing animals down, and killing them. That's when I'd start feeling bad. I'd think, "The Lord is going to hold me accountable for enjoying the hunt and the kill, rather than thinking of it as a job to do and something that had to be done."

Robert G. Cary,[3] Army infantryman, Vietnam

I was twenty-three when I went into the army and I always wondered what it would be like to kill a person. But I found out that when you are under stress and a person is trying to kill you, it isn't that hard to shoot back. Another thing that helped is that I never knew if I killed anybody or not. I've never been close enough that I've actually had to kill somebody standing in front of me.

After a year of someone trying to kill you all the time, it gets to the point where you build up a little resistance against thinking of them as your brothers or as other human beings. You just think of them as the enemy and you need to kill them.

Kim Farnworth,[4] Marine Special Forces, Vietnam

It boils down to basic survival. There's a very, very

---

[3] Robert G. Cary was born 12 October 1943 in Tucson, Arizona. After graduating from Livermore Joint Union High School in California, Cary attended San Joaquin Delta Junior College in Stockton, California, where he earned an associate's degree in industrial arts. In 1964 he joined the LDS church. He then studied for a year at Brigham Young University before going to Vietnam in 1967. After his military service, Cary was supervisor of the Games Center at Brigham Young University. He is now the assistant manager of Outdoors Unlimited at BYU.

[4] Kim L. Farnworth was born 28 April 1947 in Salt Lake City, Utah. He attended Orem (Utah) High School and married

fine line that a person has to cross to actually kill. And during three tours of Nam, I never saw anyone who, when first presented with that problem, didn't hesitate, because it's an area that you have to cross over into. Once you have crossed that fine line, it's easy. It actually becomes excitement. And, in fact, I wrote home saying I thought I was losing my mind because in a way, it actually had become a game. It sounds sick, but we'd make bets on which way their hats would fly when we shot them.

Neil Workman, Marine radio operator, World War II

We came to what they call a "pill box." It was kind of like a basement house with dirt and plants growing over the top of it. There were windows and slots on the side of it, and those inside could shoot out from these slots. We were pinned down by their firing and couldn't get past it. We had one fellow who had a flamethrower on his back. Those flamethrowers would throw a flame 150 feet. We had riflemen concentrate their shooting at the slots on a corner of this building, and while the Japanese were ducking, the guy with the flamethrower was able to get close to that corner. He put the nozzle of the flame thrower in through the opening and turned it on. This turned the building into big oven. The Japanese started running out the other side of the building, and he stood and turned the flame thrower on them. They counted about fifty bodies. You can't imagine what happens to a person burned with a flamethrower until you see it. It burned their clothes, singed everything, turned them black—it was like a weenie roast, if you have ever seen a weenie that was burnt to a crisp. The bodies actually bloated up and split open just like an overcooked weenie.

---

before leaving for Vietnam at the age of eighteen. His wife gave birth to a son while Farnworth was in Vietnam. Farnworth works at Geneva Steel in Orem.

But the thing that's amazing is how we reacted to this scene. We didn't even think about it. You'd think that we would've been sickened by it, but it was just a day's work. We just stepped over them and went on.

Clyde Everett Weeks, Jr.,[5] Marine infantryman, World War II

It was something that we didn't want to do but had to do. We were committed to doing it. We were prepared to pay whatever price we had to: to win the war, to establish peace in this country, hopefully for all time. I was called upon to fire a rifle, to throw hand grenades, to use the implements of war. I wasn't really fighting against people. I was fighting against an invasion, against the idea of having our country overrun by a foreign power, having our freedoms taken away. Many young men and women participated in the war. They were very committed to that ideal. They weren't bloodthirsty. They weren't killers. They weren't people who wanted to kill other people. We'd been attacked. We'd been threatened. Our very way of life had been threatened. I just participated and wanted to do my part to make sure that the things we treasured and enjoyed as Americans would be preserved.

Kirk T. Waldron, Air Force pilot, Vietnam

If you believe in God, and you believe in your country and you believe basically in what you are doing and you recognize the scriptural admonition that there's a necessity at times for war, horrid though it is, then you realize that you can do it. How did Abraham feel when he was told

---

[5] Clyde Everett Weeks, Jr., was born 18 November 1925 in Manila, Philippines. Weeks graduated from Provo (Utah) High School and studied journalism and business administration at the University of Utah before joining the U.S. Marines Corps as a single seventeen-year-old. He was editor of the *Orem-Geneva Times* for five years after his World War II experience. Weeks is currently the postmaster of the Orem, Utah, post office.

to sacrifice his son? He was willing to do it. I think that kind of an upbringing, where we have scriptural support and spiritual assurance that there's a meaning and a purpose behind it, allows us to understand that while certainly not desirable or pleasant, killing is at times justifiable, even necessary.

Michael R. Johnson, Marine infantryman, Vietnam

Killing had to be automatic. If it had to be done, if the situation came up, you just had to pull the trigger and worry about your head later. Better to react quickly than to be thinking, "Gosh, I'm lying here with all these bullets in me. I wish I'd pulled the trigger first." You knew they didn't care, they were going to shoot you—no qualms, no questions asked. It was a matter of just doing it and worrying about the psychological side of it later.

Jerry L. Jensen, Army Special Forces, Korea and Vietnam

The idea of having to kill people always bothered me. My church leaders told me that when serving in the military under orders we wouldn't be held accountable for what we had to do when under orders. Leaders of the nations truly would. But I really feel that there are some officers who'll also pay for what they did.

Don G. Andrews,[6] Army helicopter pilot, Vietnam

I remember an incident where I was having break-

---

[6] Donald George Andrews was born 5 December 1932 in Miami, Florida. After graduating from Miami High School, Andrews earned a bachelor's degree in psychology at the University of Florida. Parents to a six-year-old son, Andrews and his wife had a daughter in 1964 while he was on his first tour of duty in Vietnam. In June 1966, just over a year before he began his second tour, Andrews, an Episcopalian, joined the LDS church. He was a professor of military science at Brigham Young University from 1977 to 1980. He retired as a colonel from the U.S. Army and is now vice-president of Rocky Mountain Helicopters.

fast with an infantry commander very early in the morning as we were going out on a mission that day. Someone came in and told him that one of his young officers, whom I'm sure he admired and respected, had just been killed. They had the enemy surrounded in a tunnel. This colonel went on the radio and said, loud and clear, "I want no prisoners!" Well, that's against the Geneva code. But I understood where he was coming from, and I understood the emotions he was feeling. I certainly don't subscribe to violating the Geneva code, but I also understand the realities of war and what this guy was going through at the time he said that. Probably the next day he wouldn't have said it.

Jerry L. Jensen, Army Special Forces, Korea and Vietnam

I probably will have to answer for some of the people I've killed. I can very honestly say that when I killed them I wasn't thinking of mom, home, apple pie, and country. My thought was, "I'm going to kill you, you S.O.B., before you kill me." I've done it many times. I know that I'll have to someday stand before that bar of judgment and they'll say, "Why did you do this?" I think I can answer that if it was in error, it was because the information we had was wrong.

Michael Terry,[7] Army infantryman, Vietnam

When we were walking through My Lai, there were a few villagers running here and there. A lot of them were still trying to get out of the area. There were dead people lying all over the place. I must have seen fifty or sixty that

---

[7] Michael B. Terry was born 24 January 1947 in Salt Lake City, Utah. He graduated from Orem (Utah) High School and then studied math at Brigham Young University for three semesters before beginning his military preparation. As a single twenty-year-old, Terry went to Vietnam. Terry is a concrete contractor in Orem, Utah.

day. I assumed a lot of these people were killed by the artillery and helicopter gunfire.

We noticed some people lying in a ditch. It looked like a couple of them might still be alive. I mean, half their heads were missing, and things like that. It was pretty gory. Both me and the guy I was with had the feeling that we just had to make sure they were dead, because in our minds, they were goners no matter what. So we just made sure they were dead. There were two or three of them.

I've thought back on that many times, and I've always thought that what we did was the thing to do, and there was no doubt about it. I was living right, and I just went by how I felt. So I don't feel bad about it. I just felt, and still do, that it was the right thing to do.

C. Grant Ash, Army Air Forces bombardier, World War II

I think you kind of conditioned yourself by saying, "All I'm knocking out are the ships and the docks. I only want to set fire to the warehouses. I want to destroy their ability to produce gasoline. I really am not going to kill many people" — at least those were the kinds of thoughts that were in my mind. In Ploesti, Romania, I hoped that they were smart enough to get down in their bomb shelters when they knew we were coming. It was never very personal; the actual thought of the killing was quite remote. I didn't have any trouble putting distance between me and anybody I might kill.

Walter H. Speidel, German Army Afrika Korps, World War II

In the summer of 1938 our school arranged for youth camps or youth exchange programs locally with the Frenth. A people-to-people approach showed us we had much in common with the young French, who were all about fifteen or sixteen years old. We were aware that a war was imminent, but everybody hoped and wished that it could be avoided. Yet we realized that eventually we might have to fight each other.

I could never feel any hatred towards anybody, French, British, or American, probably because I was so well acquainted with the language, history, and culture of those countries and I knew personally so many LDS missionaries and the young French people.

C. Grant Ash, Army Air Forces bombardier, World War II

We aren't taught to hate people. I think our philosophy was always, "Those poor deluded Germans and that bounder of a Hitler." We hated Hitler, we could talk ourselves into that, but we had a hard time talking ourselves into hating the soldiers we were fighting. I'm sure ground troops that faced those guns every day and were shelled back and forth had to learn to shoot first in order to save their lives. But as one flying in airplanes in the Army Air Forces, when I was being shot at, I was saying, "They are shooting up at me, but wait until they receive all of these bombs down there." It was a game more than any real hatred. I never dropped a bomb in hatred.

Danny L. Foote, Marine artillery, Vietnam

It was like the enemy wasn't a real person to me. Communists, or at least their philosophy, deny the reality of God, and looking at it from that point of view, it became more of a religious thing with me than anything else. It became a battle of good against evil, not against people.

Jerry L. Jensen, Army Special Forces, Korea and Vietnam

You can't kill anyone you consider your equal. You have to psych yourself up that they are less than human, that they are animals and they deserve to die. I, like everyone else, pretty much felt that way. It was pretty easy to do that, especially when you saw the atrocities they committed. We'd go into a village and find the village chief and his wife hanging with their bellies split open; instances where they'd cut open their bellies, filled them full of rice, sewed them back up, and let the sun kill them. You'd come into an area or a village that had been friendly and find them beheaded, or find one of your patrols with their heads missing. It didn't take too long before you were quite willing to go out and hunt them down as wild tigers.

Ron Fernstedt, Marine infantryman, Vietnam

As far as killing goes, when you are fighting in a war, you don't kill human beings; you kill Gooks, Dinks, Slopes, and Krauts in the last big war. You dehumanize the enemy to the point where it's just like hunting deer. It becomes a game, so it's no big thing. And I was very good at this game. I was one of the best—at least that's what everybody told me. Men would volunteer to go on patrols if I was leading them. They'd volunteer because they knew I wouldn't waste them and they knew that with me in charge they had a chance of coming back—a really good chance. We always had fun. I don't know who made up that stupid rule that you can't have fun when you are playing guns, because you can.

You have to remember that for us, Vietnam was like an extended hunting trip. Uncle Sam was giving us all the ammunition we needed, all the food, and paying us good money to boot. I was making almost two hundred dollars a month, and for combat pay, that isn't bad. We didn't care. We were a bunch of jocks out just having a ball.

Danny L. Foote, Marine artillery, Vietnam

With the knowledge I had that death is no more than walking through another door, the whole idea of death didn't really mean a lot because in reality there was no death. People were just being sent off to other places. So the idea of death meant very little to me then and it means very little to me now.

The act of killing someone in that environment doesn't bother me today simply because, like I say, there's no death; we all continue to live. We all go on to the Great Judge and to our reward, whatever that might be. It will all come out in the end, and whether this individual went on to his reward or I did, we all continue to live, so all I did was send the guy on vacation.

J. Tom Kallunki,[8] Army infantryman and press officer, Vietnam

We got into a fire fight. I know that I shot one of the Viet Cong, one of the fellows in the black pajamas, who was shooting at us. I thought, "I wonder if this guy has a family? Where are they? What's his story?" We'd just talked to some prisoners we'd captured. I thought, "What a waste." But, frankly, I didn't feel guilty that I'd taken his life. That was never in my mind. Maybe it was the training, maybe it was the fact that I'd read scriptures that indicated people died when they opposed the freedom of others.

David L. Evans, Army infantryman, World War II

We were attacking. We'd kept them pinned down as we moved up the ridge, and as I got up near the crest of the ridge a German suddenly came up out of a hole. We'd had a few come up like this. They'd see us attacking, and they'd just been plastered with so much artillery and machine gun fire that they always came out with their arms in the air. But this one came out with a rifle and swung it toward me. I was down on my knee. Before he could aim I fired, and it was point blank. There was no way I could possibly miss.

---

[8] J. Thomas Kallunki was born 15 September 1936 in Portland, Oregon. Kallunki graduated from Bell (California) High School and attended Cal Poly College in San Luis Obispo, California, where he studied journalism. Before he left on his first tour of duty in Vietnam (1965-66), Kallunki and his wife had a two-year-old son. During that first tour a daughter was born. He went on a second tour of duty in 1968-69. Kallunki, who joined the LDS church in 1955, retired as a major from the U.S. Army. His last assignment was professor of military science at Brigham Young University from 1980 to 1983. He is currently the assistant director of student leadership development at Brigham Young University.

There was a sudden look of surprise on his face. I guess he was thinking the same way I always had: "Everybody else will get it, but it won't happen to me." He suddenly knew that it had, and it was too late. I shot him right in the face. That was a face I didn't get over for a long time—that sudden surprise, just total amazement.

Lynn Packer, Army broadcaster, Vietnam

The war made me cynical. I became even more cynical about the lack of church support for servicemen in helping them deal with the choices they have to make. I'd been taught it isn't good to kill unless you have a good reason to do it. And I thought Vietnam was really borderline. I didn't hold it against troops who went out and killed Vietnamese, saying, "Everybody understands the rules, it's a war. We can get shot and we can shoot back." But that isn't good. I don't know what I would've done had I been told to go out and shoot. It would've been a tough moral call for me to make.

As I was doing stories about Vietnam, I got press clippings from the military information service about all the stories that were being done by major U.S. media. While I was in the middle of doing stories over a several-day period, I learned that there had been two Mormons who had participated in the My Lai massacre.[9] It's one of the few times I cried in Vietnam. I read the clipping and hung my head. I'd been doing all these stories and wondering, "Had I been there, would I have had the courage to walk between the women and children and the machine gunners and say, 'This is wrong'?" I was really upset. We are taught that when we are tempted with taking drugs or with having an extramarital affair or with doing something wrong, we should

---

[9]LDS church members Greg Olsen and Michael Terry served in units involved in the My Lai incident. Both men say they did not participate in the brutalization of My Lai's civilian population.

know the right thing to do. To me it's borderline to shoot someone on the battlefield, but it isn't a borderline question at all if it's women and kids in a ditch.

It left me cynical and bitter about that, wondering why you even go to church if you don't know how to use the instruction you get there. I don't know what I would've done. I hope that had I been thrust into a situation like that, at least I wouldn't have done it, and better yet, I would've stood up for what I knew to be right. But it just shows you how far the war had gone wrong.

Neil Workman, Marine radio operator, World War II

This bothered me: we crossed the island and pushed the Japanese to the coast at the other end of the island. A lot of them were trying to surrender, but we'd been ordered to take no prisoners. On another island about five miles or so away, the Japanese had started to swim out into the ocean. For some of the guys around us, well, it was just like shooting ducks. It was target practice to see who they could hit swimming out in the water. Our guys shot them even though they were helpless in the water. It was just a contest to see who could shoot the most.

Grant Warren, Air Force rescue, Vietnam

I never dealt with the possibility of being killed. My ego was so big that I never even gave it a thought. I guess that's the truth. I just never gave it a thought. If you did, you couldn't dwell on it. I think that's why a lot of guys went in for alcohol and drugs. For me it was almost the attitude that an athlete gets, that "I can take this guy one on one and there's nothing he can do about it." I had the attitude of, "I can handle this and nothing is going to happen to me." Honestly, I never gave it a thought.

Ron Fernstedt, Marine infantryman, Vietnam

Two things about dying: dying only happens to the other guy; and I'd been promised that I couldn't be killed as long as I was living the gospel, and so I never even worried about dying. Never had any fear of it. Most of us were more afraid of being injured so badly we couldn't do things we enjoyed.

Dennis E. Holden, Marine infantryman, Vietnam

There were times when you'd be walking across rice paddies and the enemy would ambush you. You'd see your buddies fall. You'd ask yourself, "Why did that person die and I didn't?" I'd find myself saying, "God must have something planned for me or I would also have been killed." Your faith in God helped soften your fear of death.

Ted L. Weaver, Army Air Forces bomber pilot, World War II

All we could do is realize that our number could come up and then put it in the back of our minds and go ahead and do our job. That was the attitude I took. Most of my crew took the same attitude. However, my copilot had the foreboding that he'd never come back. That was the way he talked and acted. Whether that contributed to the fact that he did not, I don't know.

Kirk T. Waldron, Air Force pilot, Vietnam

Before I ever went or before I even volunteered to go to Vietnam, I considered the possibility that I might not come home. I discussed that with my wife. You don't spend too much time on ideas like that because basically they are morbid and depressing and it isn't pleasant to think about them. On the other hand, you are realistic enough that you have your affairs in order and your estate planned so that you at least in some relatively modest way have your family taken care of in case it does happen. When you know your training has been good and you feel like you are a professional and you do your job well, then you enhance your odds for survival.

Dennis E. Holden, Marine infantryman, Vietnam

Fear of death is a unique feeling that causes you to perform at abnormally high levels. Once you got over there, you came to realize that all of your instincts had to be accentuated if you were to survive. I was on patrols where 50 percent of my patrol was killed; it became clear very early that you couldn't take anything for granted. Every noise, every movement, every visual sign, and every gut feeling had to be taken seriously or you could die. I found that the careless were more likely to die than those who developed their survival instincts. I found myself detecting noises that the normal person never recognizes. Movements in the jungle never went unnoticed and every smell meant something to me. At times it seemed like I'd developed eyes in the back of my head. I know that I wouldn't be here today if my survival instincts hadn't been so strong.

John A. Duff, Army helicopter pilot, Vietnam

I lost many friends. I had a very good friend who was a platoon commander with me in the Cobra helicopters, and he had his ship blown up in midair, hit with an incendiary round, got it right in the fuel tank and blew it up, just gone. But that wasn't the only one. A lot of people did that, but you couldn't let yourself get into it. If you did that, your emotions would be so tight you'd flip, and that did happen to a lot of people. Friends died, but you didn't think about them being there. You just simply boxed up his stuff and took care of it and sent it back to his family and wrote a letter to them.

Hyde L. Taylor, Army Airborne, Vietnam

I think it's just like when you go to a funeral where they always say, "Remember, this is just this part of our eternal life, and we've got something else." That really comes home if you think about it. I thought about it a lot and I remember telling a lot of people that when they were afraid. I went through that several times.

I can remember the night a young man in our unit was killed. That night about eight of us descended into a bomb crater and stayed there. I remember telling two young men who were pretty scared about that belief. I guess I preached to them, but I didn't mean to. I think it was a calming influence.

George E. Morse, Army Special Forces, Vietnam

When I got shot through the chest, I passed out. Then I came back to my senses a little bit. I didn't hurt very much, but I realized that I'd been shot when I saw the blood. Something happened at that time so that I never worried about dying. I don't know whether I was already dead or whether I was dreaming or what happened. I was lying in a ditch, and I turned around and started firing. I'd gotten to the point where I hurt so bad and I didn't know if anybody was coming to help. I didn't want to live anymore. I closed my eyes and the next thing I knew I was standing, and it was cool and nice. I wasn't hurt and I didn't have any scars. I looked down and saw myself lying on the ground. I turned around and saw light. There was a path that I walked up with a little stream running alongside it, kind of like I was in the mountains. It was nice; it was like a summer day. I walked up over a ridge where I met my grandfather, who had died. He said, "You have to go back. You haven't finished what you were sent to do." I looked at him, turned around and walked back down the ridge. The next thing I knew I was hurting again.

Danny L. Foote, Marine artillery, Vietnam

I'd get very angry whenever an American — whether I knew him or not — got killed. It sounds like it's kind of a contradictory feeling, but you know the guy is young and he's got a family and you think about the hurt and the pain back home — it just made me very angry. I'd just want to lash out. Yet at the same time you kind of had to have a balance in how you dealt with it. So, I'd get angry and I'd feel emotional and sorry for this guy, but then I'd think, "Well, better him than me." As far as bonds, it goes back to the fact that you never really knew any of these guys for longer than six months to a year, so although you call yourselves friends, close friends, it isn't like somebody you grew up with or went to school with. You could kind of remove yourself from the situation a little bit.

Albert E. Haines, Army infantryman, World War II

I suppose one of the first jarring incidents of combat was to have a sniper take the life of one of my assistant squad leaders, a sergeant. He was virtually right alongside me. We'd gone through a brick factory and we were out the other side scouting the way to get through the village. Through a little crack or crevice in the door a sniper zeroed in and shot him near the heart. He lasted maybe a minute, a minute and a half, with me at his side and maybe a couple of others. That was when the war really became quite real. Death was part of it; not only death, but there were a thousand casualties a day — the sergeant is no longer here, he's "there." He was still a responsibility to me. Although he was dead, he was a living responsibility because he had a wife. As platoon leader, I'd censored his mail. I knew of his family, so, although he was dead, he lived on.

Lincoln R. Whitaker, Army infantryman, World War II

Our daily routine was guard duty in a foxhole if we weren't fighting. We had two men to a hole and we stood guard all night. They'd set rifles up and shoot right at the foxholes all during the night. We were to keep a guard open. One of the two men would have to keep guard all night. We had to take turns.

During one of these periods one of my best buddies, a man from Ohio, was shot right between the eyes at midnight while he was on duty and I was down in the bottom of the hole sleeping. I didn't hear anything. It must have been a small thud, because it when right through his helmet and right into his head. He was killed instantly. When I woke up the man was dead. I didn't know how he died until daylight.

Hyde L. Taylor, Army Airborne, Vietnam

I had a young man named Barton. I'll never forget his name. We were in Dak To. Our battalion and another battalion had come in contact with the enemy brigade or regiment. It was really tough terrain, a lot of heavy jungle. I remember a lot of big rocks and a lot of huge rivers up there.

We had contact off and on with the enemy, and we were doing pretty good. We came on the trail where the enemy had laid communication lines, wire. So we knew that if they'd laid wire, it was a pretty good-sized unit. Small units didn't set up those kinds of communications. Sure enough, we got into them, and they put up quite a fight.

We were a pretty small unit. As we were trying to wait and get some help from another company and calling for artillery, I warned everybody, "Don't go near the trail." The jungle was pretty thick and we were pretty safe as long as we were down in the jungle. They could fire at us, but it was a pretty slim chance that anybody would get hit. We could see fire, but we never did see anybody. There was a lot of firing going on.

I repeated, "Whatever you do, don't go on that trail." The only open area was this trail. This young man Barton just stepped across the trail and they shot him right in the head right in front of me. He dropped right at my feet and fell back in the jungle. I remembered saying just minutes before, "Don't anyone go near that trail." It was kind of sad, something that stays with you for a long time. I'll always remember him.

Lincoln R. Whitaker, Army infantryman, World War II

On one occasion we were advancing to take some high ground. This was farm country. I had a Lieutenant

Westover at my side. We were walking toward this high ground and all of a sudden machine gun fire opened up. He said, "Where's that machine gun fire coming from?"

I said, "I don't know exactly, but I think it's coming from that haystack up there."

He said, "We've got to knock that out." So we started toward it. The machine gun fire was directed mostly at the other members of our squad. It didn't appear to be directed at us, so we moved on.

As we were moving up there we discovered that the haystack wasn't a haystack at all. It was a tank and it had an eighty-eight-millimeter cannon mounted on it. They saw the bars on the lieutenant walking within a few feet of me and fired a round from the cannon directly at him. One explosion and he was totally destroyed. There wasn't anything left of him. After that battle was over they asked me to go out and identify the spot where he was and see if we could pick up any of his identification. We found his wallet, his identification, his dog tags, and a few pounds of flesh that wasn't recognizable as being from him or anyone else.

Michael Terry, Army infantryman, Vietnam

Once a guy alongside me got shot. He was a flower-child type kid, a draftee, and he hated being over there. He didn't want to have to kill somebody. It was really tough on him. He got shot, a clean shot right through the leg. He was bawling and laughing at the same time. He knew he'd be sent back home, and he was so glad.

Albert E. Haines, Army infantryman, World War II

One little vignette concerns a platoon sergeant that initiated me in the Hurtgen Forest. While we were making our way toward Frenzeberg through artillery and mortar fire, he took a shrapnel wound in the leg. He stood straight up and said, "I've got mine. Goodbye." He turned and walked away, not waiting for the battle to cool or for evacuation.

He'd seen so many of his people killed. After months and months of combat duty (Africa, Normandy, Hurtgen), he was just looking (and I didn't fault him) for a safe-conduct pass back to the United States. He didn't care how seriously he was wounded, as long as he wasn't killed. He stood straight up in that battlefield and said, "I've got mine," and he walked off the field.

five
OVER THERE
★ ★ ★

David L. Gardner,[1] Army communications, Vietnam

It was October 1971 when I first flew over the Republic of South Vietnam, and I was not prepared for what I saw. The entire landscape was thoroughly pock-marked with craters—the result of constant bombing. Having never been exposed to war or its effects, I was not sure what to expect, but the true reality of where I was now set in. I thought, "This can't be true. I am really here."

That first night I lay awake. We had giant fans in our hooch to try to keep the heat and the flies off of us. I

---

[1] David L. Gardner was born 27 December 1945 in Fillmore, Utah. Gardner graduated from Boise (Idaho) High School, earned an associate's degree in interpersonal communications from Boise Junior College (now Boise State University), and attended Brigham Young University for a year before leaving for Vietnam at the age of twenty-six. He and his wife had a four-month-old daughter at the time. Gardner was the founding chairman of the Vietnam Era Veterans Memorial Committee for the state of Utah. The Utah Vietnam Veterans Memorial was dedicated in October 1989 in Salt Lake City. Gardner, a human services specialist, is currently employed by Mountainlands Association of Governments in Provo, Utah.

don't think I slept more than forty-five minutes with all of the B-52 bombing going on only fifteen miles away. The bombs would shake my hooch, and there were helicopters and planes flying over all night long.

J. Tom Kallunki, infantryman and Army press officer, Vietnam

I frankly had read a lot of books about war and seen a lot of movies and enjoyed them. I always enjoyed the World War II movies. But there's no glory in war, not a bit. I got some ribbons and all of that kind of stuff that they give you in the military. But there's absolutely no glory at all in war.

You narrow down your perspective to the things that are really important in life. I probably did the most meaningful scripture reading, praying, and everything else during that period of time, because your life is very uncomplicated. I know that's why many soldiers, even though the threat was there that they could get killed, extended their tours in Vietnam: because it was an uncomplicated life.

George L. Adams, Army wheeled-vehicle mechanic, Vietnam

The times that were extremely hard for most of us in Vietnam were holidays. Christmas, New Year's, Thanksgiving—I think those were the hardest days I had while I was in Vietnam. Those are the times normally spent with family and loved ones doing things that are fun. There was no slow down for Christmas, New Year's, or Thanksgiving. As a matter of fact, there was more of a threat of an attack during those times because the Vietnamese knew that an awful lot of our people would tend to be inebriated during those times. We frequently experienced rocket attacks on holidays. One night during one of those situations we had rockets land right in our containment area. One rocket landed about 150 feet away from the hooch I slept in and totally destroyed two two-and-one-half ton trucks. Another rocket hit a hooch of marines and killed three of them.

Kirk T. Waldron, Air Force pilot, Vietnam

When I left for Vietnam, I had a small tape recorder and I saw that my wife had one. It wasn't the cassette type; it was the old reel type. We could send tapes home for regular postage, and it didn't cost much to do it, so I sent two or three tapes home each week. We had an arrangement about family home evening. We scheduled specific lessons or topics on certain dates, and I'd tape a contribution for that lesson and then send it in the mail. Although the kids didn't have their dad physically present, they had his voice and could say, "Yes, that's my dad. I remember what he looks like, and now he's talking to me."

Robert M. Detweiler, Air Force pilot, Vietnam

I was in Vietnam when Jane Fonda went to North Vietnam, and they published the pictures of her sitting on one of the anti-aircraft gun mounts with all of the soldiers around her smiling. I tell you, that really had an effect on the guys. They were so mad they would've gladly shot Jane Fonda. You would've had to draw straws to see who would get to shoot her.

Kim Farnworth, Marine Special Forces, Vietnam

In the particular group I was in, the survival rate wasn't too high. In fact, there were eighteen of us that went over there together in early 1965; there were six of us left when our discharge date came up. So you get hardened; you don't allow yourself to get close to anyone. All of these young recruits would come over and they were scared. They wanted to get close, they wanted to show pictures. I didn't want to see it. I didn't want to know you. Just do your job and stay out of my face.

Wayne A. Warr, Army infantryman, Vietnam

To experience stress as a group binds that group very closely together, so we were real tight. Racial boundaries broke down in that environment completely because we were us and they were them. I think I enjoyed the cohesiveness and the friendship. It isn't something we carried any further. For instance, when someone left the unit he was forgotten. If someone was killed or wounded he was forgotten, but those who stayed were very close and you could depend on each other. I liked that. I enjoyed that aspect.

Chris Velasquez, Navy combat photographer, Vietnam

I remember I came into Cam Ranh Bay after being in the field for eleven months, and I started to go to the club. All I could think of was a big thick steak. I told one of the guys and he said, "When do you want to go?"

I said, "Tonight."

He said, "Not tonight—it's black night."

I said, "What do you mean black night?"

He said, "Sunday and Monday are for Latinos, Tuesday and Wednesday are for the blacks, Thursday and Friday are for the Anglos, and Saturday is an open day." That was really quite disheartening. My own guys were isolating themselves.

Wayne A. Warr, Army infantryman, Vietnam

For me to carry everything I needed, or that they said I needed, would weigh about 130 pounds. We carried about eighty or so pounds of our own equipment and then we'd add to that mortar rounds and claymore mines. The mortar round weighed about twelve pounds, and sometimes we had two mortar rounds and extra ammunition. I carried four hundred rounds in ammo pouches around my body and then carried another four hundred to six hundred rounds in my rucksack. All that adds up to a lot of weight. If anything, we had so much that it was a physical ordeal just to transport your own equipment. It was a hindrance because most of the time you really didn't need it. When you did need it, you were glad to have it, but you might go for a month, sometimes two months, and never use it; never need to fire a round, just walk, carry it, dig, sleep, cover your hole, load it up on your back, and walk again.

David L. Evans, Army infantryman, World War II

The cold was the worst enemy of all. There was just no way to get away from it. If you went into a house, you still couldn't build a fire. There was nothing to burn to begin with, and the minute Germans saw smoke coming out of a chimney they just leveled the house. So we froze all day and then tried to find someplace out of the wind at night so that we could get warm.

Many of our men had trench foot and could hardly walk. What would happen was that your feet would get wet and if you could change your socks into dry socks everyday, you were okay. But if you didn't have any change of socks and, of course, there was no way of changing your boots, your feet would just stay wet and cold the whole time and finally they'd just start to rot. Sometimes gangrene would set in and guys would lose their feet completely.

I kept a pair of socks in my pocket, another pair of socks tucked inside my shirt, and then every night I took off the one pair and played musical socks with them, so I always had some dry ones. It was mostly the guys in the rifle companies who, for one reason or another, would end up giving a jacket to somebody else with their socks in it and forget about it. Then they'd put on an overcoat and suddenly realize they didn't have a change of socks. Also, a lot of them didn't believe anybody. They were told what could happen if they didn't keep changing the socks and they waited until it did happen and realized it was true.

Danny L. Foote, Marine artillery, Vietnam

It would get real wet from time to time, and whole villages would flood out. This was real hard on the Vietnamese, and, not being acclimated to that, it affected us even more. Our skin would start to rot and fall off. There were several times where I went for six months at a time where various parts of my body felt like they were rotting off. There wasn't a whole lot I could do about it; I just had to live with it.

Grant Warren, Air Force rescue, Vietnam

I hated the heat, the stench. In the little towns in Thailand, they ran the sewer system underneath the sidewalks. They had rotten papaya all over. It's hot all of the time. In the winter it would still be eighty-five to ninety-five degrees—in the middle of the winter! It was really funny; the Thais used to wear sweaters and it would be eighty-five degrees. We'd say, "Why do you have that sweater on, Coup?"

"It cold, it winter."

"Really, all right, gotcha."

Martin B. Hickman, Army infantryman, World War II

If I remember correctly, we didn't have a shower from the ninth of November until the twenty-first of January. As a result we began to develop what are called scabies, which are infestations of the hair follicles, so that you have these little scabs, particularly on the legs. What saved us is that we were cold all of the time, so the body stench wasn't so bad. But the feeling of being dirty and the incrustation and scabies on the legs were constant. When we finally got a shower it was in a barn-like affair, and the water was only lukewarm, but it felt like a luxury to get soap and water and wash away that three-month accumulation of crud.

During an attack just after Thanksgiving, our task was to take a village, a strong point. We marched towards the point of attack all day long through the woods, through these ravines. It was raining all day; we were wet and miserable. About four o'clock in the afternoon, and of course this is in winter in Europe and it's dark at that time, we hit the village and were driven back. We dug in and waited for daylight to renew our attack; we dug in on the side of a hill.

Our first foxhole was mistakenly dug in at the bottom of a rock face, and the water just drained down the face of the rock and into our hole. We got out and dug a new one. We threw our ammo and mortar bags in the bottom and sat on them all night. I was in the hole with a kid named Burntz. I put my head on his shoulder, and he put his on mine, and there we were all night with our raincoats pulled over us.

We woke up in the morning and we were just miserable: cold, wet, tired, so discouraged, despondent. We climbed up to the top of the hill waiting for daylight to come to renew the attack; while we were there some C-rations[2] were brought up. We took the wooden boxes and broke them up, and despite the fact that we were on the front lines we made a fire out of them. I put my hands down by the fire from that box and we began to get warm. I could feel the warmth, and my combat boots drying out; the warm feeling started in my thighs, as I stood close to the fire, and spread through my whole body. There was a remarkable lift and change in spirit that came with the spread of warmth. My attitude and feeling changed. I said to myself, "I'm not going to just lie down and die. I'm willing to give it one more try."

David R. Lyon, Army artillery, World War II, Korea, and Vietnam

It was a painful experience seeing, as the winter wore on, blood in the snow from the feet of people who had tried to return to caches of food they'd left in the ground. Having been unable to survive otherwise, they risked going between the combative forces to get food for their families.

---

[2] Combat rations, usually canned meals for use in the field. Sometimes called K-rations.

Michael R. Johnson, Marine infantryman, Vietnam

The days at times were just endless. You just became a zombie. You might be up all day, sleep for an hour maybe during a night ambush, come back in, run a day patrol, come back that night, run an ambush, and maybe sleep the next morning for three hours before breakfast.

All that time there was no fighting. You were just constantly walking in the heat, getting wet, leeches would be sucking on you and mosquitos eating you alive, but you'd have nothing to show for it. There would be nothing to release the tension and that curiosity about "Where in the heck are these people that we are fighting?" It was kind of a weird place.

If it rained every hour for sixteen days or something, you were out in it all the time. Your skin is one constant puckered up, wrinkled, split-open mess. You have sores all over your arms from the cuts on the elephant grass. You are just a walking, festering sore all the time. You sort of hypnotize yourself with thoughts of home. I used to sing Christmas carols. I didn't care when it was. I'd just sing Christmas carols in my head and walk along eating my chocolate. I used to trade my cigarettes for other people's chocolate and just eat all the chocolate they had.

We'd come off patrols at four or five in the afternoon shot, just worn out. We'd maybe lie down for an hour, get up and eat supper, which wasn't much, and then go on watch on the perimeter of the compound on the hill. Then there were these big bunkers you'd have to spend a night in every once in a while. You'd have to watch in a bunker all night long after coming in from a day patrol. You'd lie all night long in this bunker trying to keep the rats and the leeches and the cockroaches off you and then trying to stay awake to guard the perimeter. You'd become very bored, and yet there was always that underlying tension: "When is something going to happen," and, "Will I be awake when it happens?"

George E. Morse, Army Special Forces, Vietnam

When I was in the base camp, I first saw how poor the people in Vietnam were. We were fairly close to the landfill. The garbage would be taken out to the landfill and dumped. There were hundreds of kids there. They looked just like flies. The kids were fighting each other for the food we'd thrown away off our plates. That really touched me. I just had to turn away. I thought, "My hell, here I've been feeling sorry for myself, but at least in America I could go do something to earn money and be respectable, yet look at those poor kids. I wonder where their parents are."

I learned that some of their parents had been killed. Those kids were just trying to survive. I was really sick. I couldn't understand why we had so much and these kids were being left to starve to death. It used to just aggravate me because there was really nothing I could do. When I went into a South Vietnamese city, I'd see people living in cardboard boxes or under pieces of tin, picking ants or some other kind of bugs off the ground and eating them.

Timothy Hoyt Bowers-Irons, chaplain, World War II and Korea

I remember at the depot, they had a big wire fence built around the garbage cans because the local people would come in and scavenge them. One of the cooks told me about this and I went myself to see it. We'd all go out to this big wash or depression and the GIs would take these fifty-gallon gas drums with the top cut out and put the slop in there. I must say this for the cooks; they tried to help. They put the clean food in one and the gunk and garbage in another. It was all against the law, but they tried to help the Italian people get it and then they'd just dump it out, and these Italians—I've seen them—would take a gallon bucket in each hand and just walk up and hold it up, getting those two buckets full of garbage while the stuff ran all over them, stuff like that. I don't blame the GIs for trying to help. I did everything to encourage it, although it was strictly against Army regulations. What are you going to do? People are starving. They weren't getting it for their hogs and their dogs. They were getting it for themselves. So it's true, they were our enemies and they quit when things got tough. But they were still human beings.

Here again, there's one thing I have to say about the American soldier. He may have been immoral a lot of times and not perfect in many ways, but the vast majority of them would come through and try to help out. They tried to be humanitarian and help. I mean they weren't pious about it or religious about it or anything. Some of them may have been. By and large, though, if you tell them there's a bad deal and you needed some help, they'd come to the rescue. So I had a good feeling about our American soldiers. Some of them were the scum of the earth, of course, as they are in every Army, but most of them meant well and did pretty well.

Hyde L. Taylor, Army Airborne, Vietnam

The civilian communities are just pawns. I always felt bad for the poor guy who only had his rice crop every year to live for. He didn't have a whole lot to look forward to except his rice crop. As soon as the kids were old enough to wade in the rice paddy, they'd plant rice. Their whole life centered around that. Then all of a sudden a war breaks out in their rice paddy and two or three of his family are killed. He isn't a communist. He isn't anything. He's just there. He doesn't care about politics. He doesn't care about religion. He doesn't care about a whole lot of things. He just cares about surviving. Then all of a sudden he loses half of his family and the rice paddy in a war in which he has no idea of what's going on. That's injustice.

David R. Lyon, Army artillery, World War II, Korea, and Vietnam

I remember the uncertainty of the natives. They had little paper flags and they weren't quite sure who was coming or who was going in the ebb and flow of combat. You can imagine how taxing it would be for a farmer who only wanted to survive to know whether to put out a flag for North Korea or South Korea, for the United States, for the United Nations, or in later cases when the Chinese came in, for China. I've seen little paper flags of all of those nations fluttering from farm houses as the farmers nervously watched the ebb and flow of people going back and forth in their area. They didn't want to appear to be hostile to the wrong forces or supportive of the wrong forces at the wrong moment.

Ray T. Matheny, Army Air Forces flight engineer and gunner, World War II

After our plane was shot down and I parachuted out, it was a relief to be on the ground, even if it was ice. I lay there wondering why I was alive and why most of the other crew members had to die. While I was still in a numb state a boy about ten-years old came running up to me saying, *"Venga con me, Venga con me."*

At first I couldn't understand, and then I repeated his words, *"Ven conmigo."* It was Italian and I was repeating Spanish. The Spanish street of my boyhood in Watts, California, came through clearly: "Come with me." "He's trying to help me," I thought.

I struggled out of the water and sat on the ice. I motioned for the Italian boy to help with my parachute harness as my hands were still without feeling and felt like stubs. The boy was afraid to come close enough to help. I struggled. It was so frustrating for me. My brain was commanding my hands to unlatch the buckles of the parachute harness, but the hands only responded by making clumsy motions towards them, then glanced off the cold metal without grasping.

Minutes passed and the boy kept calling to me to hurry, but I gave up and lay down on the ice. I was thankful the wind wasn't blowing.

The boy uttered something and scampered off. A few minutes later a farmer, tall and worn and about sixty years old, came cautiously towards me. I sat up and it soon became obvious to him that I was harmless.

The farmer got me to my feet. I made motions about the harness but he didn't know what to do. Finally my hands started to work. Painfully I pulled the buckle and was free of the parachute harness. The farmer then motioned me to come with him. I took a step and fell. The old farmer helped me up and put his arm around my back and under my left shoulder. I put my right arm across his back and hung onto his shoulder.

My right leg wouldn't bear any weight. I'd hobbled with the farmer a few hundred yards when we came to a farm house. The roof of the house and grounds were littered with pieces of my airplane. Two teenage girls were outside the house. I made the old man stop and I pulled a comb out of my left top pocket and combed my hair. I felt improper in front of these girls and had to look better. They were pointing to the airplane debris on the roof and to me saying, "*Kaput, Kaput.*" I didn't understand German at the time but the meaning was clear.

The old German and I were met by a small group of people from the village. The villagers were mostly women, small children, and a few old men. A young man came riding a bicycle. He was designated as spokesman by the group. In barely existent English he told me that he was a school teacher. He kept asking me something that sounded like, "Where are your ears?"

A terrible fear went through me as I thought, "Good Lord, I lost my ears." I quickly touched my ears but no blood was on my fingers.

Finally after several tries I heard, "What are your years?"

I responded that I was eighteen. I must have looked sixteen and hardly a warrior. The school teacher asked if I needed medical aid and pointed to the gash on my forehead that had left a small stream of blood running down my face. By that time I was feeling better and able to stand on my own. I said that I was only bruised and cut and would be all right.

Everyone was so friendly that I couldn't believe this was the enemy. I asked if this was Denmark, thinking that our track took us on a more northerly course than planned. The young school teacher said in a friendly way, "*Nein, hier ist Deutschland.*" I was disappointed to say the least.

I was taken to a farm house next to the road about in the center of the village. There I was turned over to an Italian non-commissioned officer. He took me inside and pushed a chair next to a cast-iron pot-bellied stove. I sat down and absorbed the welcome heat from the stove.

The Italian wore an Italian Army uniform that was in need of replacement rather than repair. We talked much throughout the afternoon in Italian and Spanish. He was wounded in the hip in Africa and disabled. The Germans used broken down men like him and also young boys to help in farm communities.

The farm house was spacious and comfortable. The little pot-bellied stove warmed a sitting room that joined the dining room. The German family was friendly toward me, offering a woolen blanket, a cloth to stop the bleeding of my forehead, and food. I couldn't take the mid-day meal. It was just too soon after the trauma of the morning. The children, a boy about seven and a girl about eight, played Old Maid with special cards for the game. They spoke to me, as did their mother, a pleasant looking woman. The family Christmas tree was still up with sparse decorations mostly made of colored paper. Photographs of men in military uniforms explained their absence.

In about two hours the Italian non-com took me outside to meet Arnold Nelsen, the right waist-gunner. It was he whose parachute I'd seen below drifting to the north. His right shoulder was displaced at a funny angle and he was in much pain.

"How did you get out of that spinning bomber?" I asked.

He replied, "The centrifugal force was so great that the ball turret was torn from its mount, leaving a big hole. I crawled out the hole."

I felt stunned and could hardly talk.

Arnold was in poor shape and the Germans took him to another house to await transport to a hospital. He was taken away in less than an hour and I never saw him again.

The Italian and I passed the afternoon hours talking and going over my "escape" kit. The kit was small enough to fit in my upper left flight suit pocket. It contained silk maps of Germany and France, twenty-seven thousand francs, seventeen thousand Reichsmarks, a D-bar of concentrated hard chocolate, pep pills, and a small hack saw. The Italian soldier was totally unsympathetic with Hitler and "his" war. I burned the maps in the stove, knowing that I wasn't going to find my way to Denmark, and to the legendary escape ship. I was hurt and could hardly walk in my clumsy flight boots and flight suit. Escape at that moment wasn't even a remote possibility.

How ironic it was that on my last leave in the States I'd seen a movie with Errol Flynn playing the role of a pilot shot down over Germany. He'd escaped, of course, had blown up an ammunition train, and had made all of the connections underground, making it back to England in high adventure. To try to escape in the dead of winter dressed in strange clothes using this little "escape" kit seemed ludicrous.

I offered the Italian the Reichsmarks, but he was afraid to take them. I ate the D-bar and put the hack saw blade in a pencil slot in my flight suit. The Italian made no motion to remove it.

Supper was served about 6:00 p.m. and I was invited by the young mother to sit at the table. The Italian wasn't invited and excused himself to go outside and smoke a cigarette I'd given him. The little family was polite to me, passing the boiled potatoes, carrots, and cabbage, indicating that

I should take all that I wanted. There was heavy dark bread and butter. The meal was topped off by a slice of carrot cake for everyone. I refused the ersatz coffee.

So this is the enemy I've so carefully been taught to hate, I thought. I felt confused. The enemy looked so different from the air. Flying at twenty-five thousand feet, the enemy was impersonal, or if personified he was Hitler, shouting and waving his arms in a frenzy of madness. Sometimes the enemy was conjured as long rows of uniformed men without faces, their legs swinging stiffly in unison and arms stretched out in a salute to an idol. The enemy had been painted for me as ruthless, monocled beasts without passions, feelings, families, occupations, or religion. They were cast as automatons never doing good, only evil. They all looked alike and carried the symbol of death in their faces.

Suddenly the enemy had faces worn by toil and exposure, faces of love and concern, faces of compassion — even for a stranger, even for the enemy — and faces of fear. The father of the children, the husband of their mother, the brother, the uncle — all were in uniform. Were these the men behind the engines in Me-109s, did they man the antiaircraft batteries? "Who was the enemy?" I asked myself. I was to find the answer to this during the next year and a half.

Timothy Hoyt Bowers-Irons, chaplain, World War II and Korea

Just after I got my commission in Italy, I had a couple of days, so I just walked around Naples. I met this bright-eyed little guy, somewhat dilapidated in appearance, but wide awake. I don't think he would've been more than eight or nine years old. Of course, this was when the war was still hot. He said, "Hey soldier, you want a drink? You want to eat? You want a woman?" I said, "Heavens, no!"

He said, "What's the matter?" I said, "Well, I just don't want one." Of course, this sounds simpler than it was because his English was very GI, broken and obscene. Here I am, a young chaplain receiving this kind of proposal. Then he told me about how beautiful his mother was. I guess it's true that when a man finds a woman with a couple of kids, she's apt to go to considerable lengths to keep them fed. This wasn't uncommon at all. Little boys would pimp for their mothers and for their sisters. This is one of the things that's tragic about war.

Grant Warren, Air Force rescue, Vietnam

I saw all of these people living in poverty and filth and in total disregard for any kind of a conscience. Fathers selling their daughters for prostitution. Nine, ten, and eleven-year-old kids selling marijuana cigarettes soaked in opium outside the main gate of the base. No sense of conscience. There was poverty, stench, sewer systems in the streets. We were there saying as a nation that we were trying to protect those people's freedom. Frankly, I didn't see any freedom. All I saw was poverty, waste, sickness, prostitution, drug soliciting, and all of that. I began to wonder after that.

David L. Evans, Army infantryman, World War II

The Germans had a mail plane that kept trying to fly in, and we'd shoot it down every time it came over. They really wanted to get their mail, so they finally convinced us that it was just mail in the plane. We made them a deal that if they'd let us hook up to their electric generators, we'd let their plane through. That worked out, so we had lights and they got their mail.

Lincoln R. Whitaker, Army infantryman, World War II

Routinely, as a matter of survival, you always had to be on the alert for airplanes that might come over your positions. We had an airplane every night that would come over and drop one bomb. We called him "Bed Check Charlie." He'd come at eleven o'clock every night. You could set your clock by him. When we'd hear old "Bed Check

Charlie" coming, we'd hit the holes and try to stay down. We really didn't have too many people wounded as a result of that particular airplane.

George L. Adams, Army wheeled-vehicle mechanic, Vietnam

I was sitting on perimeter guard one afternoon and as I was watching out over the area that I was to cover, I could see some helicopters flying in low. There were three of them in formation. They were the old-style helicopters that the South Vietnamese were using. There was something hanging below one of the helicopters, but I couldn't tell what it was. As they flew across the fields and the trees, the object would occasionally hit something that was sticking up and bounce off of it. I was having trouble determining what it was. When they got up almost to my area I could see that it was a Vietnamese woman. The South Vietnamese soldiers must have had her up in the helicopter to question her about being a North Vietnamese sympathizer. They evidently got the information they wanted out of her and had pushed her out of the helicopter with a rope around her neck and hung her. They let her hang below the helicopter to bounce against whatever she happened to hit where everyone could see her as a warning, as if to say, "This is what we do to sympathizers." When I saw her it gave me some really deep thoughts and concerns: "Why are we here? If this is the type of people we are fighting to protect, why are we here?"

David R. Lyon, Army artillery, World War II, Korea, and Vietnam

The South Korean police who were endeavoring to maintain control of the populace and keep them out of harm's way when we were making our way north were guilty of such excesses of cruelty as are difficult for someone having grown up in our culture to comprehend. It's an accepted practice in such an environment that you try to get information about where people are and what they may have seen that will be of value to protect your forces and to provide them with such advantages. That intelligence, when assessed, can make your judgments better.

In one instance of cruelty on the part of the Republic of Korea (ROK) police, they with the assistance of a Korean interpreter and an American intelligence officer were asking some of the people as they crossed in a particular area what they'd seen and where. The frightened people were saying nothing. They didn't know anything, they hadn't seen anything. This was an acceptable position for them to take. The ROK policemen, embarrassed by the fact that none of the people were saying anything, took a stick about three inches in diameter and about six or seven feet long and with it broke the back of one of the peasants. In his spasm of anguish, with his back broken, the ROK policeman was still saying, "What did you see?" The man was saying, "Tell me what you wanted me to see." He'd give any answer that was wanted.

In another instance the ROK policeman gouged out an eyeball of one of the peasants passing by because they didn't seem to be responsive to questions that were being asked. It's painful to realize that many people who are innocent get caught up in the grinding machine of war and are innocently ground into pain, suffering, and death.

Robert M. Detweiler, Air Force pilot, Vietnam

I flew a group of marines down to Hué during the Tet offensive. The North Vietnamese had held that city for a matter of weeks and then they were finally beaten out and forced to retreat back home. But while they held the city, they'd gathered up all of the people who could read, any elected official, nurse, doctor, even firemen, and shot every one of them. There were about nine hundred of the local Vietnamese executed just in that area around the base.

Lincoln R. Whitaker, Army infantryman, World War II

On several occasions we found that Russian soldiers had crossed the river and gone into German homes and taken young boys out and cut their hands off and raped and mutilated the women.

This was quite a chore to protect against. At that point we felt that more than anything, we were protecting the Germans from the Russian soldiers, who were very adamant about the German people because they'd suffered tremendous losses in Leningrad and Stalingrad at the hands of the Germans. They were out to get even with the Germans because of the atrocities that had been committed upon them. So we had this element to contend with too.

Robert G. Cary, Army infantryman, Vietnam

I think there was a lot of hate by the Americans for the Vietnamese. The South Vietnamese looked exactly like the North Vietnamese, so the Americans didn't know who was guilty and often treated them all the same. Sometimes I think the Americans would go into villages and ransack them just to be mean. If they wanted something, they'd take it.

I've also seen Americans mutilate bodies. Vietnamese wear gold rings and have gold in their teeth, and I've seen Americans pry the gold out of their teeth and cut fingers off to get rings from bodies that had been swollen. For me, it was hard enough to kill, let alone mutilate someone.

E. Leroy Gunnell, Air Force pilot, Vietnam

The South Vietnamese I met I grew to love. They were a very fine people—very family oriented, concerned about the same things I was concerned about—and I just had to assume the North Vietnamese were a lot the same way but had come under a different ideology. So I had no particular hatred or ill feeling toward them. It was a political conflict that we were caught in over there. And because the governments were engaging in war, we were there to support the government we belonged to.

Danny L. Foote, Marine artillery, Vietnam

I think the worst thing was working with Vietnamese and knowing Vietnamese who had an honest desire and a dream for their country, and then leaving them in the lurch, not being able to do any more for them. I feel we never finished what we wanted to accomplish over there; I feel the war is still not over. The absolute worst part for me was seeing the frustration in the eyes of the Vietnamese. The people I became close to had a look in their eyes that said, "Don't desert me."

I had to look back and somehow communicate, "I don't want to, but I've got no control over the situation." I would've given my life over there—I would've done anything.

When I first got there, people were saying to me that the Vietnamese customs and a lot of their habits were extremely offensive, and so initially I thought, "Boy, I'm in the land of yo-yos." But as I got to know them and got to know why they did what they did and began to know them as a people that were striving to eke out a life just like we are, they became individuals and they became real people. And it became really frustrating when things went to pieces.

Wayne A. Warr, Army infantryman, Vietnam

I always felt, and so did most of us, that had the enemy been as well equipped as we were, they would've whipped us easily, because they were well disciplined, they were very well trained, and they had a cause and we did not. We were there because we were told to be there. We didn't really understand all the reasons why. It wasn't our country. We didn't have anything at stake there; they did. So it was a little terrifying at times to see how fanatical they really were and how serious they were about the business of fighting. Their whole attitude about life was so much different than ours was. We'd go to great lengths to avoid having one of our soldiers killed or wounded, but they wouldn't do that. It was nothing for them to sacrifice a great many soldiers to achieve an objective.

Michael R. Johnson, Marine infantryman, Vietnam

You never saw them, that was what was so frustrating. They attacked at night, never in the daytime. If it was daytime contact, how would you know they weren't regular Vietnamese? How do you know the people in the rice paddies didn't shoot you? They'd lay their rifles down under the water and pick up their hoes and go back to working in the rice. Just like, "We didn't do it. What do you bug us for?"

You develop a hate, a hatred of the things they could do to their own people, for one thing. The tortures they use, the things they'd do to the people in the villages who had collaborated with the marines or who had helped us in any way at all seemed so cruel to us. And yet to them, it was just life—it just had to be done and they did it. So the Oriental mind, I think, is a little different in regard to respect for life from what I learned.

You hate the enemy, you want to see them, you want to have a chance to kill them. You want to have a chance to kill them or hurt them bad, and you are never given that chance, so everything builds up in you. That's why I think some guys go crazy—when they do find people to shoot at, when they do make the decision to shoot, whether it's right or wrong, once they turn it loose, it's been building up for so long they shoot people, dogs, cows, horses, pigs, everything. I mean they'd go to a village and wipe out everything that walked. They didn't care how big it was, how small it was, whether it was a person or an animal, anything would do, and it was partly because of the frustration you felt for so long.

Wayne A. Warr, Army infantryman, Vietnam

Since the Viet Cong and the South Vietnamese essentially were the same—they were the same race, and they wore the same clothing—if they weren't carrying weapons you had to treat them as non-hostiles, and when they produced weapons you had to treat them as hostiles. So you could pass a group of people, and if they made no overt move against you, you didn't make any against them. And then you might get past them and they'd start shooting at the rear security of your unit.

Chris Velasquez, Navy combat photographer, Vietnam

On more than one occasion I saw teenage girls and boys with AK-47s. A lot of the sappers [military engineers] were teenagers. Sometimes the kids would go to the forward positions, where the machine guns are, and throw a hand grenade inside the bunker. A lot of GIs would think twice before shooting a kid, and that was their mistake.

I remember one particular base camp we were in where unusual things were happening. We'd hear an explosion at night, and the next morning we'd find a guy on watch dead. That happened two different nights. So the third night they put a sergeant out there, and at about midnight we heard the heavy report of a .50-caliber machine gun firing off a few rounds. The sergeant had challenged somebody, I guess, and they didn't answer, so he'd let go with the machine gun. Someone set off flares, but we couldn't see anything. The next morning we found a thirteen-year-old girl almost cut in two by the .50-caliber. In her hand was a grenade.

George E. Morse, Army Special Forces, Vietnam

We received some replacements that were South Vietnamese troops. We took them on a patrol with us. The fourteen of us who were Americans and the Vietnamese were doing a pretty good job. Then all of a sudden some of the Vietnamese turned and killed their own people along with six or seven of the men in my team. I don't know if they were actually North Vietnamese or what. Up until then, I never hated the Vietnamese people. I just had a job to do, the North Vietnamese were communists and the South Vietnamese were our allies. I never trusted them after that.

Kirk T. Waldron, Air Force pilot, Vietnam

There was a lot of frustration in that conflict. Vietnam is a very narrow and small country and to think that with all of the vast assembled might of the U.S. war machine, we could be held at bay by sometimes primitive weapons and forces that certainly weren't as large as ours for the most part, and whose weaponry wasn't nearly as sophisticated as ours. We controlled the air, which was one of the great keys to our successes, but still we were held at bay. You sometimes felt angry, like the politicians didn't understand what was going on, or didn't care. One just wondered why we couldn't use our power to do what we'd been trained to do and eliminate the limiting, self-imposed constraints on how we fought. Our military leaders weren't authorized to employ tactics and weapons in the way they were trained to *win* the war. To this day I maintain that if a war is worth entering, it's worth winning! It's very demoralizing to have the ability and means to win but to not be allowed to use it.

The fact is, America entered that conflict with noble intentions: to support democracy, to help a beleaguered nation, to halt the spread of communism, and we should've gone in there to win, there's no other way in my mind. If that meant using the B-52s earlier, fine, so be it. If that meant mining Hai Phong Harbor, we should've done it. If that meant attacking the heart and center of Hanoi, then we should've done it, and we should've done it years earlier. Instead, we shed our best blood, wasted resources, and lost national resolve with a crippling manacle of congressional restrictions and limitations. We felt frustration and anger at times because of the price we paid and the hippy- and Jane Fonda-type opposition to our efforts. We saw such opposition as tolerated treason.

Lynn Packer, Army broadcaster, Vietnam

There were murders going on in Vietnam in which Americans were killing each other. Most of them were committed with booby-trap hand grenades; it was called fragging. I heard there were the types of fragging where you are out on the battlefield, somebody wants you to do something that you are afraid to do or something you believe was wrong, so somebody would just lob a grenade at them. And there were people who weren't engaged in combat at all, that is, if you had a sergeant who you hated for some reason, you might frag him. This was going on all over Vietnam. Usually there would be a warning; the warning would be a smoke grenade lobbed into their sleeping quarters or hooked to the door. It's a sign you are doing something someone doesn't like and you had better change it or next time you'll be killed or seriously injured.

In Vietnam, if you felt like saluting, you did, and if you didn't, you didn't. The guy knew he'd better not make a big deal out of it or he'd get fragged.

Chris Velasquez, Navy combat photographer, Vietnam

Some of those guys would frag anyone they didn't like. Since Charlie used our weapons, when we were in a fire fight anyone could shoot anyone and no one would know the difference. If anyone tried to investigate, they couldn't prove anything.

I saw a guy get fragged one time. It was during a battle. I turned around and the guy was just lowering his weapon when the lieutenant went down. It was incredible to me, and I looked at him like I couldn't believe what he'd just done. We just looked at each other for a minute. I switched my movie camera over to my left shoulder and took out my .38 pistol, which was down by my side, and I just started to walk away—keeping my eyes on him the whole time. I was thinking, "If he starts coming towards me I'm going to unload my weapon on him." But he didn't do anything, and I didn't either. I got myself transferred out of that unit right away. I didn't want anything to do with that.

Howard A. Christy, Marine infantryman, Vietnam

This emphasis on My Lai massacre kinds of things, fragging your officers, and being all doped up on drugs and murdering civilians and so on, that's what you hear now. Well, I can tell you that for every time anything like that happened, there were times when we'd go in and give these people everything we had. We'd give them our food, we'd show love and concern for them, we'd protect them, and we truly had a heart-felt desire for them to live and be happy. And I think that if that kind of approach imbued the entire effort from the word go across the board, we would've won that war. Because love will always conquer viciousness. The Viet Cong were trying to win the conflict through terrorism. Boy, the best way to combat terrorism is with kindness and generosity and love, and we found that happening all the time, and the people would tell us that.

Danny L. Foote, Marine artillery, Vietnam

From the time you got over there, it seemed like you were in a cesspool of drugs. Drugs were available to you, and alcohol was available to you—whatever you wanted was available to you. Many times I'd be the only individual on the whole hill that wasn't at that particular point in time partaking of either drugs or alcohol, so I'd just have to go and sit and be by myself. I can remember a Christmas where I just had to go hide because everybody was either stoned or drunk, and if we got overrun, everybody would be killed, so I went and dug a hole and hid the whole night. It was a lonely experience.

David L. Evans, Army infantryman, World War II

The first time I really became aware of how callous I was getting was moving along the road when the Germans were retreating pretty fast in front of us—this was on our way to the Saar River just before we got there. They went up over the ridge and we were following after them. We took a break to stop to have something to eat, so I had K-rations in a box. There was no place to sit down and eat. We looked around for some rocks or fallen trees or something that we could sit on. There wasn't anything there except a couple of German bodies. They'd been snowed on and the snow had been blown off, and so I just went over and sat down on one of them and used that for my bench and ate my dinner and felt no qualms at all.

Lincoln R. Whitaker, Army infantryman, World War II

While we were guarding the Elbe River, the Russians were moving from Berlin towards us. We met the Russians on the opposite side of this river. The German prisoners were coming across the river in anything they could find. They wanted to surrender to us rather than to the Russians, because they knew they'd get better treatment from us than they'd get from the Russians.

That was a mass exodus. There were a lot of German men crossing that river. It was a very swift river and very treacherous. We saw them put bath tubs in the river on the opposite bank and try to float across the river. Of course the bathtub would sink and they'd drown because they couldn't swim in the river. The current was too heavy with clothes on. We saw an awful lot of men who we just couldn't help and who died in that river.

George E. Morse, Army Special Forces, Vietnam

I had a couple of friends. One got shot right through the throat at the same time I was shot through the chest. We went to a medevac hospital and then to Yokohama, Japan. I still had my stomach open so it could be drained out. There were big wire sutures down through—I could see where they'd put the stomach itself back together. My buddy couldn't talk through his mouth. He had an electronic thing that he'd put onto his neck to talk with. It didn't sound like a voice, but you could tell what he was saying. He'd come over and bother me, saying, "Come on Sergeant, let's go. We are going down to the NCO [non-commissioned officer] club."

I said, "What are we going down there for?"

He said, "We are going to get us a drink."

I said, "Hell, I can't even get out of this bed. I've got tubes all over me. I'd really like to go down there with you, but I can't."

I don't know how he did it. He had crutches. His foot was half gone, but I guess he got a wheelchair and went down to the club. He came back, and I guess he'd had more than his fair share. He came wheeling back in a wheelchair and up to my bed. He grabbed ahold of the IV bottle and shook me. He said, "Okay, are you ready to party?" He took the IV tube off the bottle and set it down. He had a bottle of what I think was vodka. He couldn't find the cork so he turned the lights on to see what he was doing. He hated to see men not have a drink with him.

The nurse just went nuts. She came in there, and then a doctor. She told the doctor, "They're crazy. One is going to kill the other. Court-martial them." They put my buddy in bed, where he went to sleep, and it was all over.

The next day I said, "Do you realize what you did last night?"

He said, "No. I went over to the NCO club. Didn't I bring you back something to drink?"

I said, "Yes. You tried to kill me. You were going to plug that bottle into my IV."

He just said, "Ah."

Neil Workman, Marine radio operator, World War II

We had a tank commander who was an "old-time" marine. He found a couple of Japanese bodies and cut their heads off. He put the heads in a five-gallon gasoline supply can and boiled them for several hours until all of the flesh came off. Then he buried them in the ground for a couple of days to get rid of the odor. Then he mounted the two skulls on top of his tank, right over the lights. Then he painted crossbones below it. Everybody thought that was great. I couldn't do that.

A lot of our men also started to collect teeth. The story went around that all Japanese had gold in their teeth, so they'd find a body and knock the teeth out to see if there was any gold in them. It was a way to show how tough they were. One guy on another island wanted to do something different, so he started to collect ears. He went around slicing the ears off the dead bodies on that island. He then strung them on a string just like a necklace. He had a string of ears that went all the way around the inside of his tent.

Chris Velasquez, Navy combat photographer, Vietnam

A lot of guys took trophies. They considered ears trophies. If they killed somebody, they'd go out and cut an ear off so they could prove it to somebody. A lot of marines and army boys did that. I saw one guy who had thirty ears on a string.

Hyde L. Taylor, Army Airborne, Vietnam

About the atrocities of the American soldier, I've been asked many times about those kinds of things. I've never seen any. Maybe I was in a very professional unit. I've never seen or even heard mention of anything that would be anywhere in that line.

In fact, once we fought a battle for three days that was terribly bloody. The unit that suffered the most casualties, one particular platoon that was kind of caught in the middle of the thing, had many dead and probably 35 percent wounded. After it was over there was a large number of prisoners, I'd say close to ninety or one hundred. There were a lot of prisoners and a lot of weapons. It was a big unit.

It just so happened that when they sent those prisoners back to be evacuated, the heavily hit platoon ended up guarding those prisoners. A lot of people said, "Somebody better get down there. They'll kill them all. They just lost their friends." I went down to see what was going on. It was handled as professionally as anything I've ever seen. In fact, they were even tending to the wounded. I don't know about those stories, I can't relate to that. I've never seen anything like that before.

I saw it on the other side and I saw a lot of it. I've been in villages, Montagnard villages way up in the mountains, where I know the North Vietnamese went through and recruited guides and took the head man of the village and cut off his head and stuck it on a pole to show the people that they meant business and that they better send the guides with them.

Ron Fernstedt, Marine infantryman, Vietnam

Before missions I'd go off by myself and have a word of prayer. My men would always check to see if I'd made commo [communication] with the Lord. After awhile, a couple of men asked if they could join me while I prayed, and then finally it became our custom to kneel in prayer before and after each mission.

It was sort of fun being the "Mad Mormon." I've had every rank from private to colonel sitting on the edge of my foxhole reading a Book of Mormon. A lot of men were looking for something over there, and there were many opportunities to talk about the church.

Once on Hill 54 we had a young Marine who claimed he was LDS. When a couple of my men caught him coming out of the local whorehouse, they beat him up. He couldn't understand that. The men I knew respected us for our beliefs.

David L. Gardner, Army communications, Vietnam

I had the opportunity of teaching the missionary discussions all over again. I was down at the Saigon branch every Wednesday for youth activities and every Sunday, almost, for church services. I had the opportunity of working with a lot of fellow LDS military people and getting to know all of the Vietnamese who had been baptized or were going to be baptized.

A friend of mine, Captain Monte Lorrigan, and I spent many hours together. He was in the presidency of the church's mission to Vietnam and later became president of the Sunday school in our branch on Tan Son Nhut Air Force base. We decided that we knew who had cut our orders to Vietnam and that being there was a great opportunity for us to strengthen our faith in our Heavenly Father. Vietnam was one of the greatest spiritual experiences of my life.

Grant Warren, Air Force rescue, Vietnam

The LDS church in Vietnam was great. It was like going from hell to heaven. You generally found an air conditioned hooch, or the base chapel that was air conditioned. We'd meet there. You would meet all sorts of guys. You would meet Marines and Special Forces Army guys and Navy guys. Most of the Navy guys were CBs (Construction Battalion). All sorts of B-52 pilots, fighter pilots, chopper pilots, and administrative guys, civilians working for the CIA, I think. We'd all come to church and some of them would prop their M-16s in the back of the base chapel and take off their helmets, and we all became Mormons, and we passed the sacrament.

There were a lot of tears shed, let me tell you, a lot of tears. You would see combat pilots who had shot down fifteen MIGs come in there and bawl. I remember this one Marine First Lieutenant came in with a Browning sawed-off shot gun (that's what he carried in the field), put it in the back, and came up and started taking the sacrament and just sobbed. Even today thinking about it makes me feel funny.

The contrast was so great. You just wanted to stay forever. Guys could never leave. All rank was dropped. We had full-bird colonels come in there who were members of the church. There would be some PFC (private first class, E-3) conducting the meeting. We'd call him President So-and-so and ask, "How are you doing?"

Some non-Mormon guys would come in there and it would blow them away. Here's this full bird colonel talking to the PFC and saying, "How are you doing, President?" They just couldn't understand that at all. They'd say, "What, uh? That's *sir*."

"Not here, he isn't."

Clyde Everett Weeks, Jr., Marine infantryman, World War II

I had one interesting experience after our return to Maui. We were getting ready to return to combat. We went back to Honolulu for about two weeks getting new supplies and getting equipped, things to use to go to Guadalcanal. I had occasion to go to church one Sunday. Not knowing the time the services started, I went to the chapel and saw the sign saying that services started at 10:00 a.m.

I got there early. As I sat there waiting for the meeting to start, I noticed the people starting to come in. It didn't take me long to realize that the service I was in was going to be a Japanese meeting, and all the people in the meeting were Japanese. The meeting started, and it was conducted in Japanese. So here I found myself, just having returned from combat against the Japanese, who were the brothers and sisters and relatives, I'm sure, of some of these same church members with whom I was sharing this sacrament meeting experience in Honolulu.

As the meeting progressed it came time for the members to share their testimonies. As they did, of course, they did it in Japanese. It was a very strange experience for me, realizing that here I was, the only Caucasian person in this meeting, and yet all these people were members of the same church that I belonged to. They shared the same testimony of the gospel that I had. Although they might have been related to those with whom I'd engaged in combat just a few days earlier, they were really not part of this war. They were not my enemies. And so, for a seventeen-year-old boy in a Marine Corps uniform in a chapel full of Japanese speaking the Japanese language, I realized that the place I found myself this Sunday had a great deal to teach me, insofar as the feelings I had against Japan and Germany.

Neil Workman, Marine radio operator, World War II

While I was in the hospital, I got active in the LDS branch. There I met a guy who, like me, had lost a leg. I started talking church to him. I took him out to church with me. The day that we got discharged from the Marines, we went to church and the next day I baptized him. That was the only time in the history of the church where a one-legged soldier baptized another one-legged soldier. It made good copy!

John A. Duff, Army helicopter pilot, Vietnam

After I joined the unit and they first understood that I was Mormon and that I didn't drink or smoke or carry on with women or anything like that, I was respected and left alone. I must say that most of the enlisted people wanted to fly in my ships or my unit. In fact, I had people coming all the time wanting to come into my platoon, because they knew the next morning when they flew with me that I was sober and that my pilots were sober and that they didn't have to worry about the pilots that they were flying with being hung over and making a dumb mistake and getting them killed that day.

Neil Workman, Marine radio operator, World War II

When I got overseas, I started gambling, playing cards. There was nothing else to do. We got into a poker game one night on New Caledonia and I won a good amount of money. The next day somebody else who was LDS talked me into going to church at the LDS branch there.

We were on leave and there wasn't anything else to do, so we went to church. This branch didn't have any hymn books, and that money I'd won was burning a hole in my pocket. So I donated it all to that branch to buy hymn books.

Timothy Hoyt Bowers-Irons, chaplain, World War II and Korea

On Corsica I heard about this boy. He was Army Air Forces, but he was in the stockade. I went down to see him. That's the lightest stockade I've ever seen. The boy, being LDS, didn't smoke, and so he got to trading his cigarettes off and found there was a ready market. Then some of his buddies started giving him their cigarettes. He'd sell them and trade them and then, I guess, some of the other people with a little higher rank learned about it, and so he got kind of a thriving little market going. He'd take a couple of barracks bags full of stuff and hitch-hike into some of those little towns up in the mountains where they were short of stuff and make a good deal. I guess several of the officers in the outfit were involved in this. Anyway, he was picked up with two barracks bags full of government loot, and, of course, he was court-martialed, but he was court martialed by his own outfit and so they gave him, I'm not sure, thirty to sixty days in the stockade. The "stockade" consisted of a fairly nice small house with one strand of barbed wire around it.

When I showed up he wasn't suffering very much, but he did wonder if the church would accept as a donation one-tenth of his money. I said, "Well, how much do you have?"

He says, "Well, I'm not sure, but I think I have in my bank account in the states around thirty thousand dollars."

I said, "Well, that poses kind of a tender question. Is it honestly gotten money?"

He said, "Well, most of it's honestly got. It's true I sold my cigarettes, but I didn't smoke them. That's where it came from. This expansion of business has only just been recently. Most of it I got through fairly honest dealings."

I considered the problem, and I didn't know who to tell him to turn all the money back to because the people who gave it to him had gotten value received and I was afraid if he tried to turn it back to the government, he'd go to jail, and I couldn't really see him spending too much time in jail. He wasn't any worse or any better than anyone else. So I said, "Well, I suggest you stop this black marketing, and when you get home you think it over and if you feel like it, you can send in an anonymous contribution to the church. If you want to send it all to them, fine, or you can give it to the Red Cross. I don't think you ought to make too much out of it because it may confirm you in being dishonest."

He was very repentant.

I don't know what happened. It kind of tickled me. There he was in that stockade, in a better house than I lived in, with one strand of barbed wire around it. I asked a couple of the officers about it afterward. I said, "You boys must have sure been mad at him."

Albert E. Haines, Army infantryman, World War II

We were pulled back from the Hurtgen Forest and went into a rest area. It was down by Verviers, Belgium. We were given our three-day passes. This is sort of an interlude, I suppose, where "Mormonism" comes into being. What do you do when you are taken off the line and you go out for recreation? Talk about painting the town red, that's what our regiment did.

The very toughest platoon leader I knew buddied with me. He sought me out and we did the town together. We had cider or something light. We heard the troops yelling, singing, drinking, and carousing, and we joined in the singing. I remember a woman singing "Roses of Picardy" in a room so filled with smoke that my eyes hurt now just thinking about it.

Then we walked the streets and reminisced. He was a rough hombre, well educated but tough. He was cold blooded; at least that's how I viewed him in a combat situation. It was interesting to have that association at that particular time. We then went back to our little hotel that they leased for us.

The next day, at his initiative, we went out on the street again. He said, "You know what would feel good?" I had the wrong idea of what he was thinking about getting me into. He turned into a place where there were some women. I thought, "I'm in trouble now." It turned out to be a beauty parlor. What he wanted was a manicure. So I had one too, and that was probably the highlight of those three days in Belgium.

W. Howard Riley,[3] Navy artillery and radio operator, World War II

One time our commanding officer came to us and said there was a dance and that he wanted some sailors to go. There were a bunch of young cadet nurses. I was one of them assigned to go to the dance. So I went and danced with a young lady, a very nice young lady. After the dance, she slipped out with me, which she wasn't supposed to do. She spent the night with me out on a park bench. She expected me to take her to a motel. She expected a lot more than she got.

Spencer J. Palmer, chaplain, Korea

Above others, the soldier whom I think of as a hero during my tour of duty in the Far East was an army private from Idaho named Bueller, a Mormon. He's my symbol of brave men, though he never knew it at the time.

Private Bueller's troopship landed at Sasebo, Japan,

---

[3] W. Howard Riley was born 18 April 1926 in Woods Cross, Utah. Before joining the U.S. Navy, he attended Davis High School in Kaysville, Utah. Riley was single when he joined the navy. Since his World War II experience, Riley, has been a linotype operator, a fruit farmer (his present occupation), and mayor of Payson, Utah.

in early March 1954, and as often happens in the military pipeline, he became separated for a time from other Latter-day Saints. I know that he felt at this time like he was merely a number on some shipping roster, one among the milling thousands.

I seriously doubt that he knew there was an LDS chaplain at Sasebo—or that I'd seen him and a swaggering group of soldiers heading out for town.

I was a little troubled about Bueller. I suspected he'd been taught right, but this was his first night in an Oriental town. He was young. He certainly wasn't with boys of his own faith. Moreover, the attitudes of most GIs at the reception center weren't conducive to morality; they'd been herded, restricted, and stifled at ports and in troopships for weeks. They'd just been paid and had money to burn. With the high *yen*-dollar exchange rate, they could literally live like kings the few days they were there. And besides, this was a curious new life and they were away from home with an entirely uncertain future. Surely this was the time, if ever, for a red-blooded man to have a fling.

Sasebo was a natural place for that breed of thinking. At night it was a city for suckers and dupes—one of dark streets, free-flowing beer, nude or half-nude cabaret shows, and endless trinket shops with second-rate merchandise. But the cheapest and easiest merchandise in town was the women. Prostitutes beckoned on every sidewalk. They were dressed in tight western clothes, powdered, daubed, and perfumed. Some were good-looking in pale light, and they proved tempting to lonesome men.

As soon as I could get free at camp I went into Sasebo to find Private Bueller. I searched for Bueller's crowd. When I couldn't spot them, I feared lest he'd surrendered to the grovelling tide or the jeers of his buddies. Then I saw him—an individual soldier standing by himself on a corner, straight and solitaire. He seemed so removed from the noisy confusion and debauchery around him that I couldn't have been more delighted.

When I asked what had happened to them, he simply explained that they'd found some women and he decided not to go along. In fact, his buddies had beaten him up when they found they couldn't convince him to join the crowd. That was all. But the words measured the courage of the man. I was proud of him.

six

CAPTIVITY

★ ★ ★

Jay R. Jensen,[1] Air Force navigator and electronic warfare officer, Vietnam

I got out of the seat, pushing as hard as I could against G-forces, and then the chute opened with a jerk. Then I started floating down. All of a sudden there was a gentle breeze blowing and it was very, very peaceful; except I could hear planes coming over, our buddies coming to try to look for us. I heard a lot of shooting at them. But except for the shooting, kind of off in the distance a little bit, it was very peaceful. Then I went into the clouds, and I thought, "Well, I better get ready, because I don't know

---

[1] Jay R. Jensen was born 29 July 1931 in Murray, Utah. He graduated from Jordan High School in Sandy, Utah. When Jensen went to Vietnam at the age of thirty-six, he had received a bachelor's degree in accounting from Brigham Young University, was married, and had three children, two daughters (ages eleven and eight) and a son (age seven). About a year before his release from a North Vietnamese prisoner of war (POW) camp, where he was interned for six years, Jensen divorced. He has since remarried, retired as a lieutenant colonel from a twenty-seven-year career with the Air Force, and worked in the aerospace industry for nine years. He is now an author and professional speaker.

what's under here. I hope we are over the ocean." And so I let my dinghy go down. You pull one lever and it releases a dinghy and it goes down, and then inflates, and floats down with you.

So I did that, and I was just going to get my radio out to radio that I was okay, when I came through the clouds. As soon as I came through the clouds, I saw that I was right over a Vietnamese village and that I'd missed the coast by about fifty feet. I was coming down right in the middle of the village. There were about five hundred people down there, and they all spotted me just as quick as I came through the clouds. It seemed like almost everyone from a one hundred-year-old woman to a ten-year-old kid had rifles. They looked like old World War I or World War II rifles. But they all had them, and they were all shooting at me. The bullets started whizzing by, and I thought, "Hey, I'm not even going to make it to the ground." I decided I'd try to fake them out, so I just acted like they hit me and just hung in my seat, kind of cock-eyed, you know, so they'd think I was hit. Almost all of the shooting stopped as soon as I did that.

Just before I hit the ground, I went through the normal procedure where I put my hand through the riser and then released the lever so the chute would collapse. Then I touched the ground and did a perfect rollover, and just as I came over they had me by both arms. I didn't even get a chance to get up, they had me that fast. It was mostly women and children, civilians. But they had rakes and hoes and rifle butts and they were all hitting me and beating me, and I didn't think I was going to be alive for very long. There was this one old lady that got my attention. She was about a 105 years old, at least. She was standing about five or six feet away and she was dipping her hand in the rice paddy and throwing mud at me.

Several days later they blindfolded me and took me over to a hill, and as we were coming up, all the people started shouting and throwing things at me. I was blindfolded with my hands tied behind me, but I could see just a little bit. I could see where I was going, so I didn't fall down. As I was coming up to the top of the hill, I heard the movie camera, so I knew they were taking movies. Just as I came up to the top of the hill, I happened to catch a glimpse. They were leading me right to where they had an American flag on the ground. Of course, some of us carried flags with us. They were trying to get me to stand on the flag, so they could get a movie picture of it. Well, just as I got to the flag, I collapsed and leaned over and kissed the flag. They didn't like that very well. All of a sudden the camera stopped, and they took me right back; they didn't do anything more. I ruined the sequence, I guess, and they were really mad.

Walter H. Speidel, German Army Afrika Korps, World War II

By May 1943 it was obvious that we wouldn't win the war in North Africa and that we'd have to give up. Rommel had left Africa weeks before the end and a Prussian general, von Arnim, had been assigned the command over the German forces in North Africa. We, the Afrika Korps headquarters company, were at that time closest to the French forces, mainly Indo-Chinese units. When we were taken prisoner, we were greeted by being whipped with leather whips by the Indo-Chinese. There were no fences around the prison camp, just an open field with guards patrolling it. Our signal corps unit was sort of camped together in that field. With five or six others I was able to get away during the night. We'd befriended the Sheik of Gafsa, and we asked the sheik to find out where the British were, and then at night we went and surrendered to the British Eighth Army. They turned us over to the American First Army, which transported us to Casablanca on the west coast of North Africa. From there we were shipped to a prisoner of war camp in the United States.

The camp was in Aliceville, a little place out in the country somewhere in Alabama. We were housed in makeshift wooden barracks, about forty people to a building. For the first month or two we built gardens around our barracks. We built a soccer field. We organized schools offering all kinds of lectures and courses, mathematics, English, and others.

Those who had artistic talents started art classes, like carving wood with razor blades. The musicians built themselves musical instruments from tin cans. I remember one soldier making a guitar from tin and wood and it didn't sound too bad. We even started an orchestra. But pretty soon the Swiss and the German Red Cross sent musical instruments, sports equipment, books, even movie projection equipment. So we had a library, and there were German feature movies.

We also had newspapers. We were able to read American newspapers, as well as a German POW newspaper that came regularly from Texas. We also had our own camp newspaper.

Later on, we were asked if we wanted to work outside the camp. Quite a few volunteered because they got tired of sitting behind barbed wire. I went out and picked cotton. That went on for a while, but the farmers didn't like it because they had to pay so much and we weren't very efficient. We got paid three dollars a week. I believe the farmers had to pay five dollars a day for us. The main thing for us was just to get out and see something different.

There were some good, friendly soldiers as guards. But there were also a few fanatics who shot at us sometimes, just trigger-happy. There were two POWs who were killed because they supposedly were trying to flee, which wasn't true. One of them went into the bushes and slept instead of working. At roll call at the end of the day he was missing. The guards got very nervous and searched around and there was a lot of shouting and then the guy woke up and came running towards us to line up. That's when he was shot and killed. I guess an army is an army wherever you go. People react the same way.

Calvin William Elton, Jr., Army Air Forces aircraft mechanic, World War II

We were disorganized for a few days, but then we became organized and were told to evacuate Manila. On 9 April 1942 we were captured. Actually what happened was that the islands fell; we weren't actually captured, the island surrendered. At that time I was a corporal and I was in charge of a ten-man defense crew that was up on the shoreline overlooking some cliffs.

The Japanese didn't realize that we were going to surrender as soon as we did because they didn't know what our situation was as far as food and ammunition were concerned. But we were very, very low in both. The night before, the eighth, we heard all kinds of explosions. The next morning we received word to come down and give up our arms. The island had surrendered. Wainwright was in charge at that time. General MacArthur had left the islands. It took about an hour from our position to go up over the mountain and then down to the central focal point. We assembled there just to put all of our guns in a large pile. The guards didn't know what they were doing and we didn't know what to do either, so we just sat down and waited.

They gave us instructions to move out. We were on the Bataan "Death March" for about five days. We'd make around twelve miles a day, but it was a matter of shuffling back and forth from one location to another.

We didn't have any food per se except when we'd stop at a compound, which was merely a fenced-in area or a building in one of the small villages, to stay overnight. They'd have big caldrons of rice cooking. If you were lucky you might get some rice. You'd have to wait in line, of course, and when your turn came to get rice, if they were out, you just didn't eat. Three of us decided to share all of the time. We shared both our water and our food, and so all three of us made it.

It took us five days to march the ninety kilometers, which is about fifty-five miles. We were put into a prison camp. I had a pair of used shoes from the quartermaster. They were a little bit large for me. They were sliding and slipping on my feet, and walking over that length of time I got blisters on the bottoms of my feet. Just about the time that we arrived at the boxcars, the blisters burst. It was difficult walking for about the next five or six days because of the rawness of my feet.

It was extremely hard to survive during that time. We were burying seventy-five to one hundred Filipinos and as many as fifty Americans a day who died from diseases, malaria, or malnutrition. For several days before we surrendered we were on half rations, mostly rice and canned salmon or canned beef. The staple was rice. A lot of times that was red rice, rice that hadn't been polished. It still had the husks on it. We were all in bad shape nutritionally.

We put all of the bodies in one common grave and then just covered them over as best we could with the dirt. During the monsoon season, as far as you dig, it would just fill up with water. You'd put the bodies in there and they'd just float, so you'd have to weigh them down as much as you could with the dirt. We did that for several days. Dysentery was quite prevalent over there, too. A lot of the people would drink water out of the Pasig River, which was polluted and dirty, but in desperation they'd drink that water. Of course, it killed several of them through contracting dysentery.

Fortunately, I didn't drink any. We had one canteen that we used sparingly. Whenever we got a chance we'd get water. If you broke out along the way to get water from a flowing well, they might shoot you or bayonet you or club you over the head with a rifle. You had to be careful of that.

Walter H. Speidel, German Army Afrika Korps, World War II

The interrogation in the United States was done in a rather relaxed manner, like asking, "How do you like America?"

Of course, we all said, "We like it, it's great. The war is over for us."

They also asked me, "How do you feel about the Russians?"

I said something like, "It's clear to me that since the Russians are communists and the United States is a capitalist country, you two will never get along. Eventually, the United States will have to fight the Russians. There's no question in my mind about that."

My interrogators were upset that I'd say this. They threatened me and said, "You maligned our allies. We have to punish you for that because this is something we can't accept." But nothing was done about it. I don't know what they asked the other German soliders, but those were just two or three of the questions I was asked.

They also asked me how I like American girls. I remember telling them, "I've never met any. Let me talk to some and then I'll tell you how I like them."

C. Grant Ash, Army Air Forces bombardier, World War II

I came up over the top of the hill. I ran and jumped over a rock wall and found myself face to face with a German hoeing his grape arbor. I almost knocked him down. I jumped over another fence and hid behind a stone wall. The old guy kept hoeing away and I thought, "A partisan, somebody who is going to help me." Pretty soon he realized that I was in a light blue uniform. It looked like I had pajamas on. It was a blue electric flying suit. Then he screamed and hollered. The long line of searchers came running back from all directions and surrounded me. Some of these civilians had clubs, some of them had pitchforks, a few of them had guns. There were two enlisted German soldiers directing them and they didn't even have weapons. They came down to me with their hands above their heads. Talking as they came. I could tell they were telling me, "Don't shoot." Then I stood up.

They stripped me of everything in my pockets. They

found all of the escape equipment. They even took my watch and white rayon scarf. Later I complained that this wasn't a government issue watch or scarf. I said, "That's my watch. You aren't going to take my watch, are you?" I didn't speak German, but I complained, "I'm an officer, he's an enlisted man." I was doing everything I could: "Officer — watch — I want my watch back!" He gave it back to me with the scarf, but not the maps and money in the escape kit.

The first thing the Germans did was put me in a small single room in solitary confinement. I was there for about two weeks. All they did was open the door and slide some food or water in on the floor. For the first two days I think I just slept, and then after that it became a little annoying. I didn't know what was going to happen and wondered if this was going to be a permanent situation. It was part of their softening-up process.

After the second day, I was being interrogated once or twice a day. The man who interrogated me was a former school teacher. He'd actually taught school in America. He was an older man in his sixties. He came in and he wanted all kinds of information. I said I'd give him name, rank, and serial number, that's all I was permitted to do.

One day he came in and he had a form all filled out for me to sign. He said, "Would you just correct any errors. We know all about it anyway, just correct any errors." I looked and was dumbfounded. They knew the signals for the day. They knew what airfield I was from. They even knew the number of the bomber I was flying in. They knew right where I'd come from. They had a fantastic amount of information.

I said, "You guys didn't do too well, but you aren't going to get anything from me but name, rank, and serial number."

He went through a little charade of, "You foolish boy, don't you know if you don't give me this information I'll have to punish you?" Sort of idle threats, he never said what he was going to do or anything. This went on for about ten days. One day he came in and he was really upset with me that I wouldn't at least sign the papers he had. Then he really threatened me with starvation.

I said, "I don't know how long I could last, I've never been hungry, really starved. I don't know how long I can take it. But until you test it out, you aren't going to get anything but name, rank, and serial number."

He went through another little charade suggesting bodily harm. I said to him, "I don't know why you people do this. I always thought you Germans were real military people, that you honored the Geneva Convention. I've always had a lot of respect for you even though we've been enemies. I've always been taught that you were gentlemen and that you'd honor the Geneva Convention. I'm very disappointed to think that you are doing these things. You are nothing like I thought you'd be." In three hours they let me out.

G. Easton Brown,[2] Army Air Forces gunner, World War II

We had one passenger with us when we got shot down over Manchuria. He was a technical sergeant that had just come over from the United States, and he wanted to fly with us on a combat mission. He got his combat.

I don't know what he did to irritate them, but they threw him in a cell with me. He was beaten so bad that he just cowered and lay over in the corner with his eyes swollen shut. He'd been hit with shrapnel from a thermite shell. He had some pieces in his buttocks and they wouldn't treat him. My hands were filthy. We didn't have an opportunity to bathe, shave, change clothes, or do anything. We just sat

---

[2] George Easton Brown was born 24 January 1917 in Lovell, Wyoming. He graduated from Lovell High School. Brown was twenty-seven years old and single when he entered World War II. Before retiring Brown was postmaster of the American Fork (Utah) post office.

on the floor. The only toilet facility was a hole in the floor. We were there in that room for one hundred days.

We soon got lice that ate all the hair off our bodies. The places where that shrapnel had gone through his skin were really festering up. I pulled those scabs off, and just like you pop worms out of a cow's back, I popped that shrapnel out of his butt. There were two fragments. Of course they had a lot of puss around them, but we didn't have any bandages or anything. He had a little bit of nylon from a parachute, but that didn't help much. He eventually healed up just fine. Boy, he was sure worked over at first. He must have just given that name, rank, and serial number routine in such a way that they made an example of him.

By the time all of this excitement died down, they took us to a school and put us in this room. We were lined up along the wall and we just sat there against it. We were all exhausted. I think about all of us fell asleep. This big, burly soldier went down the line slapping our faces with a glove to wake us up. Then they started to interrogate us. You may have heard some stories about this. The Japanese wear hob-nailed boots and take them off when they go into a building. They leave them out somewhere and wear little rubber shoes when they go inside. Those little rubber shoes were quite effective weapons.

You were supposed to give your name, rank, and serial number, which we did. They worked the pilot over, and chipped his teeth pretty bad. They'd take a perverse delight if somebody was a little bit fearful. They'd really work on them. All of this stuff about the bamboo splints under the fingernails or water torture, they didn't do that. There may have been isolated cases of it, but even though they didn't like us, they didn't hardly dare torture or kill us. They only wanted to know what we knew. They'd put us in a room and they'd take a Chinese soldier or whoever, we didn't know, and they'd beat him. They'd make this guy scream to wear you down and work on your nerves. We had this one boy on our crew whose name was Davidson. He'd just sob; he couldn't take it. Most of us could take it, you just have to grit your teeth. They worked on us like that off and on for nearly thirty days.

They put us on a truck and took us around town. We were probably the first B-29 people they had up there. They'd captured others in Japan proper, but none up in Manchuria. They took us out to a school where there were some Japanese girls. They let them throw rocks and anything they could pick up at us. We solved that by just crowding over against the Japanese guards. They got hit and soon stopped the rock throwing. They took us and put us on display. We thought they were going to kill us. They had a lot of people out there, some were boys with swords.

They got us off the truck and we all stood there with a stiff upper lip. Someone said quietly, "This is it fellows, it's been a good war, we'll see ya." They didn't kill us, but they sure scared the hell out of us. When we realized we obviously weren't going to die, we noticed our knees had become real weak. You mentally prepare yourself to die. It was quite an ordeal. Life is sweet, and even though you are being kicked around and starved, you want to hang around another day to see what's going to happen.

One time during the one hundred days, they brought in some Red Cross stationery and said we could write home. Most of the guys bought it. I said, "They'll never get out of this building." The other guys said, "Well, we're going to give it a try." I was in a rather perverse mood, because I'd been worked over the day before. On one of the sheets they gave me, I just put in big bold letters, "Horse shit. Who are you monkeys trying to kid?" I sealed it up and put my folks' address on it, and they took it. Ten minutes later this sergeant was in the room and I took another working over. Our men pretty well knew then that their letters hadn't gone very far.

They'd bring us a little ball of rice about the size of a softball twice a day. Sometimes they wouldn't bring any. The most exquisite torture I ever went through in that camp was lack of water. They did it deliberately. I'd been working for the Forest Service up in American Fork Canyon before I'd gone into the service, and I dreamt of riding up to that creek. I'd jump off my horse and run in the water. Then the water would fade away, and I'd wake up with a thick tongue and curse my captors. To a certain extent we lived on hate.

Calvin William Elton, Jr., Army Air Forces aircraft mechanic, World War II

It was Camp O'Donnell that the Japenese took us to first. They had one well and they'd work it about two hours a day. That was when you had to get your water. It was in bad shape, and lots of times they couldn't even use it for those two hours a day. You'd stand in line with any containers you had. If you were lucky enough to be up there to where you could get the water within the two hours, you got water; if you were not, you didn't get any.

Later on we moved from there to Cabanatuan. Both of these camps, O'Donnell and Cabanatuan, were Philippine constabulary encampments. They were just bamboo buildings with open air areas around them. Inside were just slats of bamboo where you slept, on both sides of the barracks all the way along. One day I got sick. I was throwing up and vomiting and running off at the bowels. I was so weak after about an hour at the latrine I couldn't even make my way back to the barracks. Arthur and two other fellows from my organization came out and got me and pooled all of the money they had and bought me a can of peaches. That was the only thing I could eat. I ate that whole can and it saved my life. I think we saved each other's lives during the period that we were over there.

We worked on the farm while we were at Cabanatuan, cultivating and planting sweet potatoes and various vegetables and also digging ditches so the monsoon rain water would run off. We did that all the time we were at Cabanatuan.

Then on 24 July 1943 they shipped me out with a group to Japan. We had to get on railroad boxcars going from Cabanatuan down to Manila. We were put so many to a boxcar—they weren't the same size boxcars that we have here in the United States, they were smaller. They'd put about eighty to one hundred men in these boxcars; you couldn't even sit down. You had to stand up as the train was going. If you had to defecate or urinate, if you were lucky enough to be close to the door, you'd do it out the door, but most people just had to go, that's all. With all the dysentery, there was a terrible mess in there.

It took several hours to get from Cabanatuan down to Manila, where we got on a ship. The ship had been used to bring horses into the Philippines for the Japanese. They hadn't cleaned it out before they put us in. Not only was the smell there, but there was no ventilation. They put us down in the hold of the ship. During the day they'd cover the holes so there was no ventilation or circulation of air in the holes. The boat was very crowded. We slept head to foot. Most POWs just wore a pair of shorts, and perspiration would roll off your body all night long.

It took about five days to get from the Philippines to Japan, where we went to the city of Ōmuta, on the island of Kyūshū. That's where we were until the end of the war. We worked in a coal mine. Several men were injured. One man broke his back. We had two that lost legs and various other injuries. We worked in ten-day periods and then we'd have a day off. It would take about ten to twelve hours a day from the time we left the camp, marched to the mine, worked and came back.

Ray T. Matheny, Army Air Forces flight engineer and gunner, World War II

According to the Geneva Convention of 1929 concerning the treatment of prisoners, a neutral inspector was supposed to visit camps and report to all governments concerned about the conditions he found. During the spring a man came to camp representing the International Red Cross from Geneva, Switzerland. This man wore a boxy, blue serge, two-piece suit that was well worn. He had a heavy knit sweater under the ill-fitting coat and he wore a fedora with a short brim. He was a man in his forties with a face reminiscent of the movie actor Frederick March. He was formally introduced to our American camp commander by German authorities. The Red Cross representative asked us how we were being treated. The Germans had just shown us a German movie that had been made in 1938. The film was a love story and had been circulated in a few places in the United States prior to our entry in the war. The movie had little value to American prisoners who understood little German. The Germans issued extra margarine, catsup, and mustard and promised vegetable seeds for a camp garden. These were obvious ploys in preparation for an inspector from Geneva. It almost worked for the Germans, but the Americans found out the Red Cross man was a German spy.

The phony Red Cross man was supposed to inspect the camp for three days, but his visit was cut short. He was in our barracks asking questions and trying to act as if he was from Geneva. He parked his bicycle outside our barracks and on a signal from Sergeant Smith, the barracks chief, two men picked up the bicycle. They carried it from barracks to barracks when the guards weren't looking toward the east end of camp. At the last barracks, under cover of several men, the bicycle was dumped into the cesspool that served several compounds.

The men in the barracks were particulary attentive to the phony Red Cross man and had drifted off into conversation that seemed to him to be valuable military information. He ate it up. But when the bicycle was safely deposited in its final resting place, signals were given and the conversation terminated. The phony Red Cross man was caught by surprise. He thought that he was doing well, and suddenly he was alone and no one would talk to him. He knew he'd been found out, but he didn't know the price he'd pay. The German spy, as he now felt fully identified, picked up his black briefcase and ran out the door. He let out an oath. "*Gott im Himmel, wo ist mein Fahrrad?* (God in heaven, where is my bicycle?)," he gasped.

His screaming quickly brought two *Wehrmacht* [German army] guards who weren't supposed to be on duty during the day and who obviously were stationed as a back-up force. The green-uniformed guards ran around the barracks looking for the bicycle while the spy retreated up the street toward the Kommandant's office. When no bicycle was found, the guards also retreated from the area.

In about half an hour a *Luftwaffe* [German air force] lieutenant came down the street flanked by six *Wehrmacht* soldiers. The lieutenant demanded that the barracks chief come outside. The lieutenant questioned Sergeant Smith about the bicycle, and of course he professed complete ignorance about it. The lieutenant got mad and motioned two of the guards to flank Sergeant Smith in a threatening way. Smith shouted back that this was abuse of a prisoner and cited the Geneva Convention. This cooled the lieutenant and he motioned the guards back. The guards were in a bad mood because they'd been on night duty, and now being called out on a quest for a bicycle seemed aggravating. The lieutenant ordered a search of our barracks, and we prisoners put up a howl and stomped our feet. The guards were ordered to get the job done and they began waving their rifles around. The prisoners slowly moved out of the barracks, but not before I saw one man struck in the stomach with a rifle butt. Since the search didn't reveal the bicycle, the lieutenant marched the six men back down the street in military cadence to the German quarters.

About an hour later another German contingent came down the street led by Oberst [Colonel] Kuntz. This time there were six *Luftwaffe* men with rifles marched by Struck, a German non-commissioned officer, and accompanied by the phony Red Cross man. Struck looked like a portly businessman, not a soldier, even when marching troops to cadence. Kuntz ordered Smith out of the barracks. Smith saluted sharply and stood at attention. Kuntz half-heartedly returned the salute, then burst out in German. Struck translated Kuntz's demand for the return of the bicycle. Smith denied any knowledge of it, which made Kuntz mad. Kuntz made a face, then composed himself and spoke calmly. Struck translated: "How could you treat a representative of the International Red Cross in such a demeaning way? He came here to inspect the camp and to bring you favors. Your men have stolen his bicycle and it must be returned."

Smith replied, "That man isn't a Red Cross representative, but a goddamned spy."

Kuntz gave an exaggerated surprised look as though he'd practiced this expression in his office. "Why, this man had proper credentials," he replied. He pulled a thick envelope from his tunic, waving it at the sergeant.

At this moment the impasse was broken by Sergeant Myers, the American camp commander, who came out of Barracks 19-A and shouted, pointing his finger at the phony Red Cross man, "This man is a German spy, sent by German Intelligence." Kuntz gave a genuine look of surprise as Sergeant Myers continued, "His name is Cedric Hoffmann, from Berlin."

This news completely broke Kuntz's countenance, and he sputtered, "The bicycle must be returned. He can't get a replacement until after the war."

At that the prisoner spectators quietly snickered. Kuntz spun on his heel, motioning Struck to come, and they marched down the company street. We'd won. The spy was exposed, his bicycle gone forever, and Kuntz couldn't punish us.

C. Grant Ash, Army Air Forces bombardier, World War II

The camp was such that the Germans allowed you a certain freedom inside restricted compounds. We were counted twice a day and fed three times a day, not much, but a little something. Sometimes they just forgot, at least they didn't come some days.

There was more neglect and boredom than there was meanness. They weren't mean or cruel to us in this camp. They were trying to abide by the Geneva Convention as best they could, at least at that stage of the war.

At this time they had probably four thousand to five thousand prisoners of war in the Sagan (now Żagań, Poland) area. They had us divided up in compounds so that they could handle us easier and we couldn't riot and push out. In fact, if you read the book, *The Great Escape*, the real "great escape" took place in the camp that I was interned in. The escape took place two or three weeks before I arrived. I remember there were big signs that said, "Escaping is no longer a sport. Anybody caught escaping will be shot." I'm sure the Germans meant exactly what the signs indicated.

Ray T. Matheny, Army Air Forces flight engineer and gunner, World War II

April 1944 was a cold month with very few sunny days. The most notable event in April began about the middle of the month with the addition to the camp of a single prisoner. The new kriegie [prisoner, from the German for "warrior"] was a staff sergeant named Johnson who had been captured by local police. Since a prison camp was nearby, the police took him to the gate. The *Wehrmacht* guards simply opened the gate, and Johnson walked in. In a few hours he disappeared, and that's when the Germans got terribly upset.

Johnson went straight to the American camp commander, Sergeant Myers. Myers already knew Johnson through his network of communications. This was the third time that Johnson had been captured, and now there was a price on his head. The Gestapo [Nazi secret police] came that afternoon and demanded that Johnson be turned over to them. Myers said he hadn't the faintest idea what they were talking about. The Gestapo men were large, well fed, and wore civilian clothes. Each had a felt hat typical of the Austrian area with short feathers stuck in the band. They wore long black overcoats, but one man had a gray tweed coat. The sight of these men brought fear into our hearts, knowing their terrible power over civil and often military matters. The Gestapo ordered all prisoners onto the compound for roll call. All prisoners were accounted for, so the man in the gray overcoat asked for a dog tag check. The guards had to read each dog tag, and prisoners had to say their numbers in German. All prisoners were still accounted for except Johnson. The Gestapo agent in gray then demanded a picture check, so we lined up again while their "mug" shots were compared to each prisoner. It was late afternoon when a tatoo check was demanded by the Gestapo agent. Johnson had a tatoo on his chest and upper left arm. Prisoners were made to strip to the waist to check for tatoos. We stood in long lines while guards and Gestapo looked for the tatoos.

After the prisoners had been checked against their photographs, dog tags, and tatoos and Johnson still hadn't been found, the Gestapo announced that we'd remain outside until Johnson was turned over to them. The Gestapo went through each barracks, tearing up bunks, throwing everything on the floor, stealing Red Cross-issued items, cigarettes, and anything they wanted. The search included the latrines, store houses, cook houses, and every building within the prisoner fences.

We prisoners remained in the cold all night while the Gestapo continued to tear things up. In the morning, three Gestapo men stood stoically dressed in long overcoats and their silly little hats. One spoke in English saying that if we didn't reveal Johnson's hiding place we'd remain outside for the remainder of the war. The prisoners didn't snicker or laugh at the Gestapo's remarks. These were men not bound by a military code of conduct. They were responsible only to Hitler and to themselves.

The kriegies withstood Gestapo harassment out on

the compound for three days before being released to the barracks. I'll never forget the final scene of five Gestapo men angrily striding across the compound to the west gate in total defeat. Not a sound could be heard from the Kriegies until the Gestapo men had marched out of sight down the hill towards Krems. Then there was a huge cry of victory.

"What happened to Johnson?" was the question being asked. At first there was no news on what happened, but later when the Kriegies were being evacuated from camp, the American camp commander gave out the story. Johnson reported to the American camp commander within minutes after being thrust into the camp by the police. Myers had been alerted and knew that Johnson would be killed by the Gestapo. Johnson had been hidden under the east shelf of the latrine for Barracks Seventeen and Nineteen, which had been prepared at the time of cleaning.

The day after the Gestapo left, we heard on BBC that President Roosevelt had sent a message to the German High Command warning them that the American government wouldn't tolerate mistreatment of prisoners and that they'd be held responsible for the events of the last three days at Stalag 17-B [German prison camp]. This news gave us great courage to know that our president took a personal interest in us. "How did the president know so soon about our plight?" we asked. "Myers must have a direct line out of here," someone replied.

Calvin William Elton, Jr., Army Air Forces aircraft mechanic, World War II

I never gave up. I never thought I wouldn't return. In fact, when we'd get in a group at night, we'd always talk about what we wanted first when we came back to the United States. Maybe it would be strawberry shortcake or pineapple up-side-down cake with cream topping or something like that. We were always thinking about a good meal that we wanted when we came back.

The food under the Japanese was a very meager,

basic rice diet. In the morning we'd have what we called "lugow." It was rice cooked in a very mushy consistency. We'd have a little bowl of that. For lunch we had another bowl of steamed rice and perhaps a vegetable of some kind or a piece of fish. Sometimes two meals a day, sometimes three meals a day, but you never knew. If you were sick you received a smaller ration than if you were well. They felt that if you were healthy and producing for them, you were entitled to the rations. If you were sick, you weren't entitled to them.

Ray T. Matheny, Army Air Forces flight engineer and gunner, World War II

The group that I belonged to bought a bread knife and sharpening stone from Smitty in Barracks 17-A. Knives of any kind were forbidden by the Germans, but we were issued a kilo loaf of dark bread two to three days a week to be divided between thirteen men. Such a loaf of bread had to be cut carefully if thirteen men were to get equal shares. Smitty from Barracks 17-A made knives out of our window latches, which the German carpenters had installed in 1919 when the barracks were built. The steel latch was about fourteen inches long and one and one-half inches wide with two slots cut through to receive locking bolts. Smitty worked the window latch down with a file he'd purchased from a German guard. A wooden handle was fitted with copper rivets, making an efficient and handsome bread knife. Since each group of partners required cutting of bread, there was at least one knife in each half end of a barracks.

I was skilled in doing fine work, and I was noted for my ability to cut things evenly. I didn't like the way the bread was being cut, and one day I grabbed the knife and began to cut slices. My cutting was careful, and I was given the job of cutting bread for our little group. When bread was issued, everyone was concerned because division of the loaf into equal parts was essential lest there be a serious argument. The new loaf was set on a table and thirteen men would gather around to inspect it, lift it, and smell it. I'd begin marking the top of the loaf with the knife into the thirteen parts, compensating in thickness where it tapered at the ends. When the marks were made, they were inspected and discussed by the men. Names were put on a piece of paper or cardboard and each assigned a number through thirteen. Each man had a number corresponding to a numbered slice of bread. These numbers were fixed, but the men shifted their names one number each time a loaf was issued so that a different slice of the loaf was received. As a fee for my skill in dividing and cutting the loaf, I had the right to scrape each side of a slice of bread. The heavy, moist bread would produce curled crumbs that would dry out in a few minutes and be lost. If the residue was scraped off immediately, it could be recovered. I had a milk can for collecting crumbs, and in two to three weeks the can could be filled. I was given the name of "Cutter," and often my services were sought among the other groups in other barracks, particularly when an odd-shaped loaf of bread was issued.

C. Grant Ash, Army Air Forces bombardier, World War II

The Red Cross parcels that we obtained had four packs of cigarettes, a can of coffee, and a can of milk—"Klim milk"—which was the first kind of dry milk. I never had any difficulty trading a can of coffee for a can of milk, so I'd have two cans of milk. You can see by just that alone how much more nourishment I was getting than some other guys. We could always trade the cigarettes for somebody else's food, someone who would rather smoke than eat.

Ray T. Matheny, Army Air Forces flight engineer and gunner, World War II

Cigarettes could buy anything but freedom at Stalag 17-B. Cigarettes were ranked according to the quality of tobacco blend. Lucky Strikes were first, with Camels second, Chesterfields third, Avalons and Old Golds shared fourth, and several other poorer brands ranked down to seventh place. The Red Cross parcels had Lucky Strikes, Camels, and Chesterfields, which provided a fair stock of money in camp. When these brands of cigarettes were plentiful, they were smoked, but during 1944 there was less smoking and more trading of these valuable items with the Germans.

Some Germans got conned by enterprising prisoners, who carefully unsealed a pack of Lucky Strikes and a pack of Avalons and exchanged them. The government tax seal wasn't put on these cigarettes, and it wasn't too difficult to open and reseal the packages without detection until the product was tried. It was common to see a German guard talking to a prisoner and the guard carefully smelling the cigarettes and sometimes looking each one over. Cigarettes became the most sought after items in camp, and they effectively were equal to money.

My first mail arrived 9 September 1944, nine months after I'd been shot down. My letters and cards home requested chocolate, gum, and cigarettes. I was greatly surprised to receive a parcel of ten cartons of Lucky Strike cigarettes. My parents had written the company about my request and they sent me the ten cartons. Somehow the package got by the Germans, and it was like passing a bag of gold coins to a prisoner. Suddenly I was rich. I immediately bought a can of Spam and a concentrated chocolate bar. I had plans to buy a radio, maybe a camera from the gambler in Barracks 17-A, and plenty of food.

It didn't take long for second thoughts that told me I couldn't keep all of those Lucky Strikes for myself. I gave away nine cartons, package by package to friends and kept one carton for myself. The carton of Lucky Strikes contained two hundred cigarettes, which I traded for about six hundred Chesterfield, Camel, Old Gold, and Avalon cigarettes. Then I hired two gamblers, bought two decks of cards, stocked them with cigarettes, and sent them out on a circuit, avoiding big games. The little syndicate that I formed was called "Deacon's Sinners," after my airplane. Within a month I expanded to four gamblers and had sufficient income to buy cans of meat when they were available. My goal was to buy a radio, which would take several hundred cigarettes of the top brands.

Calvin William Elton, Jr., Army Air Forces aircraft mechanic, World War II

I have no animosity against the Japanese at all, individually or as a country. I had good experiences with one supervisor while I was working in the mine. His name was Kimoro. He was very kind. He was a middle-aged man probably in his early forties at the time. He'd give us breaks whenever he could. He liked the Americans. When we'd sit down to have our lunch, he'd get in a conversation about old-time American movies such as Charlie Chaplin or Laurel and Hardy and other comedies he'd seen. He was very considerate about giving us as much free time as he could, but he was limited, of course.

G. Easton Brown, Army Air Forces gunner, World War II

They had a small compound built around the old house where we stayed. They had an electric fence along the top, and they'd make us stand out there in the sun sometimes. They'd march us around and line us up.

There was one boy with us who had played football for the Buckeyes at Ohio State. He was a great big, husky guy. He broke his foot while bailing out of one of the other ships that was shot down. Of course, there was no medical treatment for it, and it healed in rather a grotesque fashion. They'd line us up and he could never put his feet right. That one foot would always go off to the side.

We named one of our guards "Eleanor" because he had buck teeth like Eleanor Roosevelt. Eleanor was a sadist! He loved to hurt people. He'd stomp on this foot, sometimes until the Ohio State kid passed out from pain. The rest of us had to stand there and take it. If we showed any sign of fight, they'd put a bayonet against our stomachs and dare us to move. We'd have to take it.

We'd spend hours after these ordeals plotting how we'd slowly kill Eleanor. We had some real cruel ideas of how we'd finish Eleanor off. We drew lots and I won. The day that they came over and said, "Our countries have made peace, you are going over to this main camp and then you can go home," the guys looked at me and said, "What about Eleanor?" There was a short-handled hickory pump handle out in the yard, and I said, "This ought to do the trick." I planned on finding and killing Eleanor just as sure as I'm talking to you here now. I wasn't going to torture him as we planned, I was going to dispatch him rather fast. All the guys looked forward to it because of the punishment we'd taken through the months from him.

When we got over there, one of the first questions we asked the other prisoners was, "Where's Eleanor?"

They laughed and said, "You are too late. The Chinese got him last night." They took us and showed us Eleanor's body. It was lying next to an upright post upon which his decapitated head had been impaled. You could tell it was Eleanor. He'd run afoul of too many people, and the Chinese prisoners got him. Now, I'm very relieved that I didn't have to follow through with it.

Ray T. Matheny, Army Air Forces flight engineer and gunner, World War II

"Herman the German" was a character that could've fit in well with the Three Stooges. He was chunky and at about five feet, eight inches, he weighed in at near two hundred pounds. Herman sometimes tried to fit the Nazi role but made a terrible mess of it. He was sloppy and looked as if he would've been more at home in a blacksmith's apron than a *Wehrmacht* uniform. Herman was terribly naive and easily led into a circumstance that he couldn't salvage. Some of the men taught him "English" phrases, and he'd repeat these in a gleeful way. Herman would stomp up the barracks steps making a noise like a bull. Someone would ask, "*Was ist los* (What's the matter), *Herman?*"

Herman would reply, "*Nichts ist los, alles ist verboten!* (Nothing's the matter, everything's forbidden!)" in a joking way.

Then someone would engage him in a silly banter of words, often swearing and trying to get him to repeat them. Some "English" lessons would include, "Herman, *sprich,* 'I'm a dumb son-of-a-bitch.' " And, of course, Herman would try to repeat the phrase without asking what the German equivalent meant. After a few days, when Herman learned the phrases well, the Americans would give him some fanciful German translation. Herman would often repeat these phrases in his loud barracks-calling voice for everyone to enjoy. This game went on for over a year until there was a chance encounter between Herman and Struck, a German non-commissioned officer who spoke English, when Herman was demonstrating his newly-acquired "English."

"Big Stoop," another German, didn't spend much time in the barracks area because he feared for his life. Big Stoop was so named because he looked like that character depicted in the comic strip "Terry and the Pirates." Big Stoop's German name wasn't well-known, but he fit the comic strip characterization. He was about six feet tall and probably weighed 250 pounds. He had an enormous face with a great drooping jaw and an incredibly stupid appearance. He was an exception in the *Luftwaffe*, which was selective about who wore the blue uniform.

Before I arrived at Stalag 17-B there was an escape attempt that ended in disaster. A tunnel had been cut through from a barracks next to the fence out to a small grove of trees that was used as a cemetery. The tunnel was dug by a small group of men joined in secrecy. The break was made by digging out the last few feet of vertical shaft at the end of the tunnel. The Germans had knowledge of the tunnel, where it was to break out, and the time of the escape attempt. When the men broke out of the vertical shaft, they were greeted with searchlights and gunfire. One man lay wounded with a bullet in his leg; others were killed. Big Stoop went over to the wounded man and fired eight shots into his face despite cries for mercy heard by men in the barracks. This barbaric act labeled Big Stoop and the officers in charge that night, including the camp Kommandant. The tragic event devastated escape groups throughout the American compounds, and when I arrived every new prisoner was regarded with considerable suspicion. Hence, it took many months to establish a relationship with the escape groups. No informant was exposed while I was there, but presumably the informant continued to live in the barracks.

Another German *Wehrmacht* soldier who played a role in the escape episodes was "Abe the Mole." I never knew Abe's name, but he was so nicknamed by the Americans because of his facial and physical appearance. The name Abe was a great insult to him; he became furious when he heard it. He was portly, very short, and had a face that reminded Americans of a Jewish stereotype. Abe's job was to find tunnels by crawling under the barracks and running a steel rod into the ground, hoping to find excavations. He was always accompanied by a large German shepherd that protected him and presumably could smell out a tunnel. Abe came as unobtrusively as possible, but when he was caught crawling under a barracks, everyone would stomp on the floor with their feet close to the place where Abe was. The noise was terrible, and it disturbed not only Abe but the dog. The prisoners made Abe's life miserable with derisive names and remarks. Soon Abe could only probe for tunnels when the prisoners were kept out on the compound for an extended roll call.

Walter H. Speidel, German Army Afrika Korps, World War II

We heard and read about how the war was going. The greatest fear for us was that the Russians would come into Germany because we knew what would happen if the Russians came. We were hoping that the German generals would just open up the Western front so that the Americans and the allied forces could occupy Germany before the Russians would, but of course that didn't happen. Even after the Americans had occupied most of what is now known as East Germany, they retreated and turned the area over to the Russians, a fact which to this day is incomprehensible to the people who lived through it because it was clear to everybody that eventually there would arise some controversy between the United States and the communists.

In June 1945 there were teams of American information officers coming around with some Germans showing us films from Buchenwald, Auschwitz, and the other extermination camps. We couldn't believe it. I remember one time when they tried to force us to work in the cotton fields and we said, "No, we won't work, because according to the Geneva Convention you can't force us. It has to be voluntary." We'd volunteered before to work in the fields, but as soon as they tried to force us, we all refused.

Then we were told, "If you don't go to work as we order, we'll put you in a camp like Buchenwald." We just laughed, because at the time we hadn't seen that film. We said, "So what's wrong with Buchenwald?" We'd never heard what "Buchenwald" was. When we were shown the film we said, "We'll never believe that any German could do anything like that." However, the film seemed to be accurate, it seemed to be too horrible to be just made up like that. We eventually believed it and had to swallow the pride that no German would be able to do anything as horrible as that.

C. Grant Ash, Army Air Forces bombardier, World War II

We had our own secret radio. The Germans would find only a piece of a radio, never a whole radio. We had enough parts so that at a certain time of day at a given barracks a group of six or eight guys would assemble. They'd put a radio together with the parts they carried. They'd then receive the news broadcast from England, which was our voice in Europe and was propagandized, just like the Germans' news; not always right up to snuff. But we could then compare the English version against the German news. We knew that somewhere in between there must be something that was right.

Ray T. Matheny, Army Air Forces flight engineer and gunner, World War II

Christmas Eve 1944 was anything but a cheerful time for the Kriegies of Stalag 17-B. There were loudspeakers that the Germans rarely used posted in many areas of the camp. This night, however, they blared Christmas music that made us feel homesick all the more. The worst was when the Germans played one of Bing Crosby's latest records, with Bing crooning, "I'll be home for Christmas, if only in my dreams." Many Kriegies cried openly as they heard Bing's voice. To top off the problem of Christmas Eve, the temperature had slipped to minus thirteen degrees Fahrenheit. Christmas day I had a cheese and liver paste sandwich made from cans saved from days when such food was available.

The effects of confinement were showing on some of the kriegies, including myself. I clearly remember a kriegie who went out of his head, jumped over the west compound warning wire, and climbed about halfway up the fence when a guard in the gate tower shot him with a rifle. Two men had tried to stop the distraught man, but he had too much of a head start. The guard was cursed, shouted at, and clearly told that he and his commanding officers would pay for this deed. The guard, in a display of irrationality, waved his rifle back and forth at the men standing below, threatening them. It was nearly an hour before the man's body was removed.

After the shooting I questioned my survivability. There was a rumor that five men each week were losing their minds. These men were allegedly taken to Vienna for psychological treatment but were never heard from again. It was true that many Kriegies were affected by the shooting and by the disappearance of these men. I'd wake up in the night feeling that maybe I was going crazy and fearing that I might also try to climb the fence some day. During these lonely hours I'd listen to the noise of the barracks; men snoring, someone having a nightmare, the rats that came out after everyone had retired. I often felt as though I was losing control of my mind. I'd bite my hands to try and break a trance-like condition that I felt was coming over me. There was genuine fear that if I didn't fight off this strange sensation of losing control of my mind, I'd be carried away to a Vienna mental hospital or worse. One night I nearly went over the edge of the fearful state when a rat ran over my hand as I lay in my bunk.

C. Grant Ash, Army Air Forces bombardier, World War II

We were at Luft Stalag 3 near Sagan. We were there through Christmas. On 27 January the Russians came down toward the old Polish border on their race to Berlin. We heard them. We could hear their big guns, the artillery, for weeks.

We finally left the camp and were forced to march westward. We'd been alerted two weeks earlier that such a march may occur. They weren't going to surrender us to the Russians. They were going to keep us as barter. In the middle of the night, they marched us out. After about three days of marching, we'd just about had it. The fifteen of us had dwindled to twelve, and we were pulling a sleigh that we'd made out of a bed. It had bedding, pots and pans and the like on the sleigh. We were trudging uphill, a long stretch of a hill. I remember we were low on food. I was eating a can of margarine that had come in the Red Cross parcels. I remember thinking, "I can spoon the margarine out with my hand and eat this straight fat without getting sick." I hadn't slept for over forty-eight hours, maybe nearer to seventy-two. We were forced to march at night and rest in the cold for four or five hours in the middle of the day. We were moving rapidly because they thought the Russians were trying to capture us.

I remember at one point going up this hill. I remember saying, "I take a mouth full of margarine and it's just like putting fuel on the fire, I can feel it immediately go to my muscles to keep me moving." Then I suddenly realized that I was no longer in my body. I was floating, and there I was down there. I was floating above the whole column. I went down the column and saw all that was going on there. All of these guys that were falling out were being taken by the Germans to farm houses. They'd knock on doors and just push them in and say, "We'll be back for them. If they escape it's your neck." Then I went up the line and looked over the hill, it seemed to be all down hill from there. I then came back and went back into my body.

When I got out of the war and returned home, I talked with mother and we compared dates and notes. As near as we could tell, within a few hours of the event related above, my mother had awakened in the middle of the night very concerned for my welfare. She was so concerned that she'd gone into the bedroom and prayed for my safety, to know my whereabouts, and to ask the Lord if he'd at least tell her whether I was whole, or if I'd lost any of my arms, legs, or eyes. What was my condition? "He's alive, I know, but what condition is he in?" she pleaded with the Lord. Eventually, she fell asleep and said she had a dream. She said she was sitting at the sewing machine and I came in one door and went out another. I came in and she said, "Oh, Grant!" But I wouldn't talk to her. I stood before her and held up my hands and showed her all of my fingers and showed her that I was walking and went out the door. "Oh, Grant, come back." I did it three times. The last time I said to her, "Mother, you will just have to be content," and walked out.

She described me. I was wearing an overcoat that she'd never seen any military men wear. She described a French officer's coat that the Germans had given me as we left Luft Stalag 3. When we left it was twenty-seven to thirty degrees below zero. They'd given me this old French officer's coat that flared downward from under the arm pits because I was obviously not sufficiently dressed for the weather conditions. It was tight in the shoulders and then flared, just like a big skirt. It was all wool and went clear to my feet. It was actually a little too big for me. I had the white rayon scarf I had with me on the airplane. I'd obtained from the Red Cross an olive drab scarf. I'd sewn shredded paper between the two scarfs to make them warmer. I liked that rayon against my face. Mother described this white scarf with the olive drab side. There was no way for her to know this. I didn't have gloves, no, but I did have my good shoes on, she remembered. And, I did! She was given the privilege of knowing that I was all right.

Calvin William Elton, Jr., Army Air Forces aircraft mechanic, World War II

We awoke one morning and the Japanese camp commander told us that we didn't have to go to work. We were going to honor the dead that had died there in our camp. We marched out on the parade ground in formation, and he gave us a big spiel about being friends; he wanted to be friends with the Americans, and he hoped the Americans would be friends with the Japanese. He wanted to honor those that had died in the camp, and he wanted us to honor them. Then, he dismissed us. He said, "The rest of the day is free." We knew something was wrong. We hadn't heard definitely about the war being over or the Japanese surrendering, but we knew something was wrong.

By the middle of the day, we determined that the war had ended. We broke down the walls of the camp and took control. We went out and got food wherever we could, whether it was live on the hoof or whether it was rice or whatever, and we brought it back to camp and cooked it up and ate it on an individual basis for the first day or two. Then the Japanese started bringing in food.

After the Japanese surrendered, we stayed in camp perhaps four or five days, and then on an individual basis we just took off and boarded trains. We'd heard rumors that the Americans had established an air base on the southern part of the island, so we just got on a train and went down there and were repatriated. From there we went to Okinawa and then to the Philippines and back to the United States.

Walter H. Speidel, German Army Afrika Korps, World War II

I think it was in the middle of April 1946, almost a year after the end of the war, when we were put on an old freighter and shipped to France. We weren't told at first, but we found out later that the American government had agreed to turn over approximately 380,000 prisoners of war to France to be used in the coal mines, to rebuild bridges and roads, and to help repair the destruction of the war.

Soon after we arrived in Camp Bolbec, Le Havre, we learned that we would be given a physical examination that would determine who was fit to work in the coal mines, or on roads, or in removing mines and bombs—the duds that hadn't exploded. I went through the physical examination on the second day. I wasn't quite sure if I should go first or last into the long line for the examination, but then I figured, "The longer it takes, the more tired they get, and maybe later they'll be less attentive and might be more inclined to let people slip through. Maybe they had a certain quota to fill." So I decided to line up close to the end.

Around midnight my turn came. There was a German doctor first, then an American doctor, and finally a French doctor. The French doctor had to make the final decision. When I came to the German doctor, he read my name and said, "Speidel, do you have a brother George?"

I said, "No, George is my cousin." Which was true. He said, "George and I were in France together. We were the best of friends." He asked me, "What's wrong with you?"

I said, "I always had trouble with my heart."

He said, "Forget it. That's too easy to detect."

Then I remembered when I was sixteen or seventeen years old, while I was on the track and field team in school, I had a knee injury. It was called Schlatter disease. I think it's called Osgood-Slatter disease in the United States.

I was treated at that time and didn't have any problems at all later. I sort of grew out of it and had all but forgotten about it. I told him about that injury, and he was quite excited and said, "That's great! With bad knees they can't use you in the coal mines, and you can't work on the roads either." He wrote "Schlatter-Osgood disease," and the American doctor just smiled at me, looked at what his German colleague had written, and okayed it. Then I came to the French doctor. It was after midnight, and he'd had too much wine and by then was just a very happy person. He just looked at the piece of paper and said, "Oh, Schlatter, oh yes, good old Professor Schlatter in Heidelberg. I studied under Professor Schlatter," and he marked my sheet "unfit for work." The next day some of us lucky ones were on our way back to Germany.

Timothy Hoyt Bowers-Irons, chaplain, World War II and Korea

There was one thing that impressed me very much. I guess it convinced me I hadn't really been prepared for the war. It had always seemed to me to be kind of fair and square. They were trying to kill you and you were trying to kill them. We had a difference of opinion. I tended to discount, to some extent, the horror stories. I knew from my reading that during World War I there had been a lot of propaganda about horrible things that happened, and so on, so I sort of tended to minimize all the horror stories I'd heard.

Then we overran Nordhausen. There was a concentration camp there, and there were three thousand bodies piled up. They had this huge barn-like structure, almost as big as an aircraft hanger. They'd painted black stripes on the cement floor, just about six feet long and about three feet square and there were bodies laying on the stripes.

Some of them were empty. They were to keep track of the bodies. The Germans were very methodical, apparently. Some of the bodies had their wrists and their ankles wired together. They were still alive. Some of them were so weak that they could hardly wiggle their eyes. You didn't know they weren't dead. We lost 150 that night from malnutrition and mistreatment. These were the slave laborers. I was mad enough to kill everybody. I would've wiped out the whole German nation. You don't treat people like that, but they did. There they were. So we unwired them, but they died like flies, they were so weak. They wired them like that not just to be mean but to keep them from moving. Sometimes in death throes they'd kick around and roll out of the proper place and then you would lose track of them. How do you talk about that? You can't believe it. I can remember how I felt, and I can't believe it to this day, but it was there. I saw it. I have pictures of that pile of bodies. We used bulldozers to dig big graves, and then we carried the bodies down and laid them in there and kept track of them and covered them up with a bulldozer. What the hell are you going to do with that many bodies? It isn't like having a little funeral service.

Suddenly the war got very serious. Up until then it had been adventurous and kind of romantic. It was uncomfortable, but you were a hero and there was this great crusade in Europe. It suddenly got very grim for me.

Freeman J. Byington,[3] Army Field artillery, World War II

Our outfit overran a Nazi concentration camp called

---

[3] Freeman J. Byington was born 6 August 1912 in Lava Hot Springs, Idaho. Before entering World War II at the age of thirty-one, Byington graduated from Logan High School and studied agronomy and plant breeding at Utah State Agricultural College (now Utah State University), where he earned bachelor's and master's degrees. Before retiring, Byington worked in agricultural finance and banking and as a management analyst.

Ohrdruf near Mahlhausen, Germany. One of our squads was assigned to go into this camp and bury the dead. One of the prisoners who had escaped told us through an interpreter that the prisoners had been mostly Russians and Jews. He said that when the Nazis realized we were approaching the camp, they decided to move out and take the prisoners, mostly political, with them. Any prisoners who were too sick or crippled to travel, they just gunned down. The Nazis had lined them up and shot them, and the bodies were still lying there. As we went into the camp we opened the big warehouse and inside were dead bodies just stacked up in there by the hundreds. The Nazis had been putting them into ovens or taking them into the woods to get rid of the bodies. Most of these prisoners were nothing but skin and bones, they were so badly starved. You could understand soldiers being killed as part of the war, but this torture of prisoners was senseless. I can't imagine how one human being could do that to another. It was one of the worst things I saw in the war; I'd seen some of the destroyed cities, but these concentration camps got under my skin worse than anything else.

Eugene E. Campbell, chaplain, World War II

I found our people were very bitter toward the Germans as they moved across Germany and began liberating the concentration camps and American prisoners of war. We ran into one unit where the prisoners—fliers who had been shot down and so on—were starving skeletons. They hadn't been cared for. I guess communications had broken down as far as the German Army was concerned, so these men looked terrible. It just infuriated us. As we liberated concentration camp areas, you just felt like going out and shooting every German you saw. Some men actually lost their perspective; we had people in our division ride along in the truck and see farmers out in a field and begin taking pot shots at them. You lose a lot of sensitivity when you get into a uniform and a gun is in your hand, especially when you've seen atrocities.

Lincoln R. Whitaker, Army infantryman, World War II

We came upon Gardelegen and were digging in for the night when a man approached us and told us that he'd just escaped from a prison camp. He also said that the prison camp wasn't too far away. So we decided to go with him, and we found a prison camp that the SS troops had held. The camp was for Poles, Jews, and refugee people who hadn't been sent off to Auschwitz or some of the other murder camps.

The Germans knew we were going to overtake them. They wanted to leave, but they didn't want to leave any prisoners behind. So they went into the camp and put fresh straw on the ground throughout the barracks and told them that they'd now have nice places to sleep. Then the Germans went in there and poured kerosene all over the straw and they set it on fire. They had a machine gun set up at the gate. Many of the prisoners had forced the gate to the point where it was almost opened, but not quite. There was a pile of dead people at the gate. I'd say it was head high, at least five or six feet high.

The man that had come to us had been pushed down by his comrades trying to get out, and he'd been trampled underfoot. The machine gun overshot him, and he wasn't wounded. After the fracas, the German SS troops came back into the camp and said, "Those of you who are still alive, if you will stand up, we'll take you to the aid station or the hospital and get medical help for you." This individual could hear the shots fired, and he watched these German officers shoot the people that would indicate that they were still alive.

When we got back there with him, we found two more individuals alive who had been buried under the piles of dead bodies. The smell was so horribly bad; I just don't know how any of us stood it. We combed the camp and tried to find out if there was anyone else still alive. We unpiled those bodies that night. That was a terrible experience. We saw firsthand what the Germans were capable of doing.

We reported this man's story to the higher authorities. As a result of this experience, our company commander went and got the *Burgermeister* [mayor] of the town and asked him if he knew what was going on out at that camp. He said he didn't know. Our company commander took him out and showed him what the SS troops had done. The German people couldn't believe it themselves; at least they said so.

Our company commanders and our battalion commander charged the *Burgermeister* with the responsibility of erecting a cemetery there and keeping that cemetery forever green in memory of everyone that had suffered there. It had a sign, both in German and in English, telling the story of what happened there.

Hyde L. Taylor, Army Airborne, Vietnam

We walked into a prisoner of war camp one night quite by mistake. We just parted the grass and walked right into it; it was that well camouflaged. It must have had about forty Vietnamese prisoners of war being held by the North Vietnamese. It was way up in the mountains, made out of bamboo and thatch.

They discovered us there. The security people that were there were all North Vietnamese. They put up quite a fight. It took us all that night and the next day to get in there. One of the things they did to discourage us was to push the prisoners at us and try to get us to kill the prisoners as they pushed them ahead of them. They tried every dirty trick there was in war. We finally got some help from another unit.

We went in and rescued about thirty-five of them. I think five of the prisoners were killed. The conditions there were like you see in Nazi Germany. They had them lying in leg stocks almost starved to death. I had an interpreter with me, and he could talk to them. They had them on work parties. They'd take them out and they'd do menial labor, such as work on ambush pits or things like that. They were in just horrible condition.

They had a large latrine dug. I'd say it was about twelve feet wide and maybe twelve feet deep. They'd covered it with logs and left openings where they could use it for a latrine. As these workers would become too weak to work, their guards would just go push them in there, dead or alive. If they died, they threw them in there. If they were just too weak to work, they threw them in there. That's the kind of conditions those people were under.

We had to carry them out because there wasn't a place to land a helicopter in the area. I remember I had one over my back and his legs around me, piggy-back style. We got down to where a helicopter could land and he got off. I remember he got off and I laid him on the grass. He left a layer of skin on me. That's how emaciated the people were. Somebody told me later that about thirty-two of them lived; three of them didn't make it. None of them would've lived very much longer if we hadn't gone in there.

Maybe bringing those people out is a redeeming value. Maybe that's why we were there. Maybe those thirty-five lives wouldn't have amounted to anything, but it made you feel like you did accomplish something.

Neil Workman, Marine radio operator, World War II

We'd taken a Japanese prisoner. He looked like a fifteen-year-old kid. Our soldiers questioned him and tried to get some information out of him. He didn't even know what they were talking about. It became obvious that they couldn't get any information out of him. The captain who was in charge said, "Well, we can't be bothered with him." He just took out his pistol, put it to the prisoner's head, and shot him.

Howard A. Christy, Marine infantryman, Vietnam

On one occasion, I regret to say, I pulled my .45, cocked it, and, with my finger on the trigger, placed it to the head of a Viet Cong suspect that was brought to me during a company sweeping operation. I did so to scare him into telling anything he might know of the whereabouts of any nearby enemy.

The man just stood there and shook from head to foot in fear. But he repeatedly stated that he knew nothing. I dropped my arm, uncocked the weapon, replaced it in my holster, then instructed the interpreter to tell him that he was the bravest man I'd ever met. I am, and will always be, ashamed for having so cruelly treated him.

Eugene E. Campbell, chaplain, World War II

I've met a lot of men, men who were in combat a long time, who said, "We just learned not to bother taking prisoners in certain situations. They were just a handicap to us, and we just took them out and shot them." Those things happened. Toward the end of the war, I saw our men capture literally hundreds of German prisoners. They'd have to haul the prisoners in big trucks from the prison camps, which were sort of big barbed-wire entanglements where they took them temporarily. I saw our men rob them of their wristwatches, whatever they had on, and then force them into a truck and tell them to crowd up. The driver would start the truck and stop it in a hurry and they'd all fall forward. Then they'd stick another ten prisoners in so they were packed tight. They'd roll down those mountain roads so fast that a prisoner couldn't hope to escape. Going around in those narrow towns or narrow forests, the guys on the outside often got killed, hitting their heads against trees. Human life became pretty cheap. War is a gruesome, bloody, miserable, insensitive business.

seven
LEADERSHIP

Pat Watkins, Army Special Forces, Vietnam

We used troops that were mercenaries; they weren't Americans. We had language barriers, ethnic barriers, and a lot of problems that the average soldier in combat with people of his own country doesn't have. I was a little worried about the situation and rightly so. I had to prove myself to them. Most of these people had been fighting all their lives, and here I was, a basic novice, coming in and having to take them out on the ground and perform operations. I was more nervous trying to prove myself to the people I was out there in combat with than I was of being afraid of dying or being in combat. I wanted to prove myself.

It took about three months in combat before I started to get the attitude that I knew what I was doing and that I could keep us alive. It was a learning experience. The areas that I went through with the troops were especially the first and last parts of an operation when people tended to get lackadaisical. I constantly trained, I worked hard. It was a job to me. I considered myself a professional soldier. It wasn't like I was a draftee over there, just trying to make it back in one year. My whole mission was to keep my people alive and to keep myself alive. I really worked hard at it, and I think I had a pretty good reputation for doing so.

Werner Glen Weeks, Army helicopter pilot, Vietnam

I had quite a cross-section of personalities in my squad, and I wanted to be sure that my men were well informed. Each of my men knew I'd be fair. I ran a duty roster from the first day, and every man knew who he'd follow in performing extra duties that came. It was somewhat the luck of the draw as to where they'd be going and what they'd be doing, some duties being worse than others.

On a particular afternoon, a fellow by the name of Sheradon, a Southern fellow, quite mouthy and immature, was assigned to pull all-night duty as the "CQ" runner. CQ stood for charge of quarters, and this trainee was assigned to be the errand boy for the officer in charge. The duty ran from five o'clock at night until five o'clock in the morning. Very early that day I informed Sheradon it was his turn and told him to be sure that he was physically and mentally prepared to report for duty at 5:00 p.m.

We'd be released from daily training usually about 4:30 in the afternoon. That was when everybody would finally have the opportunity to let down. Soldiers would invariably go to the PX and buy sodas, or, more often, beer. I specifically warned Sheradon that he not get drunk before reporting for his duty. It would be quite hazardous to his career. He disregarded my instruction, and at about ten minutes to five, he came into our barracks to inform me that he was in no condition to pull duty. I said, "Sheradon, you know it's your turn. I feel it unfair to send anybody else in your place. It doesn't matter to me what condition you're in at present. I think you've made your own bed. Now go lie in it."

He walked out the door, suggesting that he'd probably be rejected for duty. Within three minutes he was back in the barracks, his mouth going quite loudly, threatening that he was going to kill me. The company commander, on seeing his condition, had restricted him to the company area for the next month and reduced his pay. Of course, all of this was my fault, which he quickly let me know.

Sheradon told me that he'd kept a live round in his gun from the rifle range, and since we were sleeping with our weapons as part of our training, he was going to use that round and put me away forever. Fortunately he was at the far end of the barracks bay. Fourteen men of the squad lay between him and me. That night as I went to my knees before going to sleep. I was very troubled, and my prayer went something like this: "Father, I'm worried. Sheradon said he's going to kill me tonight. If he comes my way, please wake me so that I might be able to defend myself." As I closed my prayer, my head hit the pillow, and I've never enjoyed a more restful night's sleep in my life. I literally slept like a baby.

As morning came, I awoke and checked myself to see that I was still alive. During that day, the greatest tribute ever paid me came to my attention. Six of the other fellows in my squad privately approached me to tell me that they'd lain awake during the night waiting for Sheradon's foot to hit the floor. They were going to give him a "blanket party" if he tried anything. A blanket party is earned in the service when someone refuses to fall in line. His contemporaries throw a blanket over his head and work him over to try and get him to conform. Fortunately Sheradon had gone to sleep quite quickly that evening.

Liem Quang Le,[1] South Vietnamese Army Airborne, Vietnam

I had command over twenty-four people. That was

---

[1] Liem Quang Le was born 21 November 1955 in Vietnam. He graduated from high school in Vietnam. Members of his family were members of the LDS church when he was serving in the army, but Le did not join until some time later. After serving in an airborne division of the Army of the Republic of Vietnam (1972-75), Le immigrated to the United States. He is employed at Geneva Steel in Orem, Utah.

really hard, especially for me. Most of the guys underneath me had a lot more experience than I. Some of them had been in for ten years. And then I came in and had to tell them what to do. Oh, it's hard. They give you one hell of a time. You know that they might even kill you some day. But you must be tough on them. One reason I like the airborne is because they rule like steel. That means if the men don't listen, I can take out my .45 and shoot them.

George E. Morse, Army Special Forces, Vietnam

I saw a lot of young officers that came to Vietnam out of officer candidate school. They'd wear their butter bars and they'd make a big thing about their rank. They'd say, "How come you guys aren't standing at attention? Get into a position of attention when you talk to an officer."

It took me a little while to understand, but I'd take those young officers over behind a tree and tell them, "Sir, let me tell you a little something about this country. You see those pretty gold bars you have on your collar? If you want to put black ones there, that's really neat, but for the rest of our sakes, would you take those damn things off? You aren't back in the states now. We understand your responsibilities. But let me tell you something, sir, until you've been here six or eight months, if we can get you to live that long, and you've been in the field with us and have let us teach you how to survive, you might get out of here as a captain." Some of them would listen to me and some of them would not. Some of them wanted to be the leader, "Forward ho," but they didn't know what they were doing and they were killed.

Robert M. Detweiler, Air Force pilot, Vietnam

When a fire broke out in the cockpit, my copilot, who was only twenty-two years old, just froze—absolute panic. He started crying and saying, "I'm too young to die." I said, "Just fly the damn airplane. I'll do everything else." But he just completely collapsed. He was trying to fly the airplane and he almost stalled it out. He was listening to that flame and he just kept saying, "I'm too young to die." I said, "Just sit there. Don't touch anything." I flew the airplane with my knee and one hand and did everything else, including calling on the radio for a steer back to the base and talking to the crew to get them ready. If you have that sort of a situation with a crew member and he happens to be in charge of the airplane, the result is you lose the whole crew. You lose a dozen guys because of the ineptness of one person.

Freeman J. Byington, Army field artillery, World War II

I ran into Generals Eisenhower, Patton, Bradley, and Walker at the Ohrdruf Concentration Camp on 12 April 1945. Sometime later during our "end run" going behind the lines down towards Austria somebody had fouled things up. Parts of the U.S. Third Army under Patton's command were crisscrossing each other. It got so bad that Patton stationed himself in the crossroads to direct traffic. It was a bit unusual to see a four-star general up there on top of a tank directing traffic.

Jerry L. Jensen, Army Special Forces, Korea and Vietnam

As a sergeant I was assigned as a liaison to an Ethiopian outfit that was in Korea. It had an American unit on both flanks. One was an Army unit, Twenty-fifth Infantry, and the other was a Marine unit on the right flank. I was assigned there because I understood the American tactics. The Chinese hit us again; that was the second mass attack I was in. This one got pretty hairy. They hit the Ethiopians full-bore, just like a pile driver. They hit us, hoping to break through and come in and encircle the two American units. They misjudged the strength of the Ethiopians; they were tough. The flank that I was on started to falter. The lieutenant that was there, an Ethiopian, was killed. The Ethiopians, much like the British, had to have an officer to tell them what to do. I just stepped in, not for any heroic reason, but strictly because I wanted to stay alive, and started commanding the Ethiopian sergeants who were there in our counterattack. It got down to hand-to-hand fighting. Before it was over I was bayonetted through the hand. We finally looked around and there were more of us than there were of them. We held.

That day, just about an hour after the battle finally subsided, the major in charge of the Ethiopian unit came up and asked who was in command of that flank. An Ethiopian sergeant who was there said, "Sergeant Jensen was." I was called forward and I met the major for the first time. He congratulated me for taking over. He said, "You should be an officer." The next day I was. They swore me in as a second lieutenant. An American general did it.

Lawrence H. Johnson, Army Air Forces bomber pilot, World War II

After we'd left the target and were well out, we headed toward China, way out away from Formosa, and then made a dog-leg turn to the left and headed down toward the Philippines. Everything was quiet on the airplane. We were flying back all by ourselves. My navigator, Kirk, was a fellow from Virginia, a very gentle, tall fellow. Kirk was deathly afraid of the copilot, Carnahan, who was an Irish type who was vindictive and could swear a blue streak. Kirk came up and stuck his head through the crawlway space there and said to me, "Larry, can I talk to you without Carnahan hearing?" And I told him sure, that Carnahan was exhausted and had gone to sleep.

He said, "I don't know where we are." And I thought he was kidding. He said, "No. I got so scared during that run, I didn't take any times at any place. I don't know where we are."

I said, "Well, can't you do a dead reckoning and give us an approximate position?"

And he said, "Well, I'll try. I'm not sure I can even do that."

We were headed back to the 90 percent Japanese-held Philippine Islands, and supposedly we were going to try to find a landing strip that was going to be blacked out, and they wouldn't dare turn on a radio beat for us or anything and wouldn't even dare talk to us because of radio silence, and we didn't even know where we were. I just said a little prayer right then, and I had the best comforting feeling come over me. I did a fast little sketch, and I put a mark on the map, and I said to Kirk, "That's were we are."

He said, "How do you know that?"

And I said, "Unless you've got something better, take us from there and see how we're doing." Never had a worry. He went right in and hit the base right on the head.

Howard A. Christy, Marine infantryman, Vietnam

I was given command of a company in the vicinity southwest of Da Nang, and I got into a dreadful mess the first day. We'd been sent into an area that I didn't know anything about. My company was in one battalion, but we were pulled and attached to another battalion — to beef it up for a large-scale search and destroy mission in the region of the southern boundary of I Corps.[2] We of the forward element exited the aircraft and established a temporary company position right at the airstrip, and I checked in by radio with my new battalion commander. We hadn't been on the ground an hour when elements of the battalion became engaged in heavy combat with a strong Viet Cong force, and we were alerted to move to the point of contact as soon as ground transportation arrived. We loaded onto amphibious tractors (amtracks) and headed north across a wide, shallow river, where we were dropped with orders to form a blocking position through which the units

---

[2] Vietnam was divided into four areas of jurisdiction or responsibility. The northern sector, known at I Corps, was the responsibility of the Marines.

engaged with the enemy could withdraw safely once contact had been broken.

We arrived in the late afternoon, and the platoons went into line and started to dig in along the river bank while elements of the battalion filtered through our position and back across the river. We were soon alone, with no machine guns, no mortars, nothing to really protect ourselves close in. Before we had a chance to prepare much more than very shallow prone holes, just at dusk we were attacked by mortar fire by an unseen Viet Cong force. I was so confused as to what was happening that my first impression was that we were receiving short-round fire from our own artillery. I got on my tactical radio net back to the battalion and called for a cease fire. The operations officer of the battalion told me that the "friendlies" weren't firing; rather, we were taking enemy fire. I should've realized that because I'd been in intelligence long enough to know that most enemy mortar attacks occurred at dusk, which is a confusing time because it's hard to tell where fire is coming from. So there we were, green and inexperienced, with no support units, taking deadly incoming fire. Within a minute or two we had two killed and seven wounded, and we hadn't even seen any enemy yet.

After the firing stopped we received orders to pull back. Then transportation arrived. It got dark, and it started to rain. It was about 8:00 p.m. when the amtracks showed up. I ordered that a careful muster be taken before we loaded up. A four-man intelligence team from the battalion that I hadn't met had been attached to me before we deployed so hectically a few hours before, and there was no one from the command post group to watch over them. We couldn't find them and I ordered a search. We were extremely vulnerable to attack as we milled around in the dark looking for those men. The amtrack commander, knowing the danger we were in, frantically enjoined me to load up immediately. I hesitated, unwilling to commit the worst thing a marine can do, which is to abandon any of his men.

I was a captain and the amtrack commander was a major, yet we both knew that I had the decision-making responsibility as the tactical unit commander on the field. Finally, knowing we were tempting fate, I ordered the withdrawal, hoping that the intelligence team was somehow with us. It was. As it turned out, the men had crawled into a corner of one of the amtracks, oblivious to the dilemma they'd caused.

Upon recrossing the river, matters became even more complicated. We received a coded radio message to immediately commence an extended patrol to a position about ten kilometers distant, where the company was to hold up and await further orders. I was dumbstruck. Again, there we were, in a hostile area cut off from viable outside support and without our organic support or weapons elements, in the dark and in the rain, with two killed and seven wounded, green, and exhausted. To be ordered out into the night in totally unfamiliar terrain known to be occupied by a dangerous enemy seemed ludicrous, if not potentially suicidal. With great trepidation I picked up the radio and asked to speak to the battalion commander personally. I was taking a desperate chance in asking a senior officer I'd never seen to rescind an order he may have given personally. The colonel came up on the net, and I explained the situation and requested withdrawal to the battalion area to reestablish control — and hopefully to be joined up with all elements of my company before once again being deployed. The colonel was very understanding and calmly ordered the company to remount the amtracks and fall back to the battalion area. I don't think he knew of the night patrol order I'd received. The order was probably concocted by an operations staff officer as a "routine" mission. By training and instinct I thought so, but I also feared that the requested withdrawal would be considered cowardly or an indication of my weakness under pressure. It was a terrible start. My first moves could well have been my last. Fortunately, the battalion commander respected my judgment — and in time that judgment proved to have been correct.

The second day was nearly as bad. We were once again ordered out on patrol, and our headquarters and weapons elements still hadn't arrived. It was obvious that, in addition to the shock of the day before, the troops were ill-prepared for combat. Both the junior officers and the men were tentative, scared. They seemed to be waiting for me to tell them every step to take. I ached for my first sergeant and gunnery sergeant, experienced men who could take some of the burden.

Suddenly I was struck with dread that the company would possibly not be able to stand up if ambushed or otherwise attacked. I knew that once bullets began to fly there would be little I could do if each subordinate commander — platoon, squad, fire team — didn't assert control in his own sector of responsibility. When firing begins, every individual must respond almost instinctively. In the initial confusion and furor of fire received and returned, the company commander can do little. It's too late to hold school. It was terrifying. I wanted to escape it so badly that I began to fantasize that were I to step on a mine or be shot — wounded sufficiently so as to be evacuated — I could honorably be relieved of the crushing pressure.

The word "honor" is very important here. The pressure experienced by a combat commander is in large part owing to a strongly felt need to perform honorably. In this case I desperately wanted to get away, but I just as desperately wanted to quit honorably. The clash of these feelings nearly unnerved me. It was the severest test of my life.

The third day the company, now at last at full strength, was again ordered on an extended patrol, this time back to the very river bank where we'd been the first afternoon. There we were ordered by battalion to hold up for a period. I ordered all hands to dig in, expecting immediate compliance. Surely no one would need to be reminded of the deadly mortar attack we'd suffered only two days before. Yet, many of the men approached the task halfheartedly. One squad leader seemingly didn't even pass the order; he and his men just stood around, smoking and talking. I was furious. I grabbed the marine, a young corporal, and shook him like a father would his child. He fell to the ground. Choked with rage I relieved him of the command of his squad and placed the next senior man in charge, completely ignoring the chain of command by first demanding compliance through the corporal's platoon commander or through the company gunnery sergeant. Those who observed must have begun to wonder. That night I wept—from exhaustion, self-doubt, rage, anguish over men who had already been killed, and dread that the company wasn't competent enough to survive the days ahead.

The company gunnery sergeant stepped in to where I was and, seeing the state I was in, quietly asked if I wished him to prepare the necessary marching orders for the next day. I said yes and thanked him for his thoughtfulness. He prepared the order, called in the unit leaders, and read them the order, explaining that it was my order but that I was sick and had asked him to deliver it for me. Although we didn't discuss the matter further, I believe that he offered to prepare the order for more than one reason. We were good enough friends that he probably sympathized with how badly I felt, but as a professional he also knew that the men might not understand the truth of the matter, that they might think of me as being a weakling, or possibly beginning to crack under the strain. Indeed, he may have wondered if I was in fact beginning to buckle. Whatever the case, it was best to keep it quiet, at least for the time being.

Then my "answer" came. The next morning I arose convinced that our collective survival depended upon me alone and that if the men didn't fear the Viet Cong enough to be disciplined and alert, they'd learn to fear me. From that moment until my relief weeks later I drove the men relentlessly and humorlessly, relieving of command or otherwise punishing any who failed to maintain strictest discipline. The approach was successful. That is, the company did survive, even to the point that it won considerable respect throughout the battalion and regiment as a disciplined and hard-fighting unit. That success, however, was gained at the cost of nearly unanimous and lasting hatred on the part of my men. Few could see the need for the discipline which at least at first was so coldly forced upon them. I think that most were too young, too sanguine about life, to see the terrible danger they were in. And none carried the burden of responsibility that so heavily weighed upon me.

Wayne A. Warr, Army infantryman, Vietnam

When we set up an ambush, it was very quiet. In some ways it was kind of a break because there was no digging, you just quietly moved in. My troops were pretty well trained and knew the routine, so there would be no talking. We'd just ease into a place and set up the ambush site the way I described it. I'd usually go out with one squad and recon the area so that I'd already know what the site looked like before I went in. I'd lead my platoon into the site because I knew the situation, and then by virtue of training I'd drop squads off and they'd set up by putting out Claymores.

My platoon could usually occupy an ambush site in about ten minutes and have no noise. We didn't want to be discovered, of course, so in that sense we were trying to hide. But there you would sit and every noise that you would hear all night long you would think, "I wonder if this is someone coming," especially if it was real dark. It was very difficult to tell. In fact, if one of my own troops happened to move around or something, he might get shot by my own folks, so it was a little nerve-racking. Often there was a lot of time to reflect during an ambush because we'd have to run several of them before we'd have some kind of activity. So it was peaceful, kind of a break, but the stress was higher.

I'd usually trigger it. As the leader of the platoon, it was usually my preference to trigger an ambush and then pull out with the theory that whatever I was ambushing would soon get reorganized. Because we were so far from our own unit, we didn't want to get caught alone, especially if they were an overwhelming force. That was one of the worries of an ambush patrol: that you were away from the larger unit. I could get myself into some real trouble, and at night I didn't really know how big of a unit we were ambushing, though sometimes I did. I remember one instance when I did know and I didn't trigger the ambush because of the size of the unit. I had twenty-two people in my platoon at that time, and I started counting heads as they came through and at thirty I chose not to trigger it. I estimated about a third of this organization went through this trail at night where I was sitting, so I didn't trigger the ambush. We sat there and watched them all go through and hoped that nobody coughed or anything.

Don G. Andrews, Army helicopter pilot, Vietnam

The morning of the assault we take off at dawn. I'm flying in what's called a command-and-control aircraft. I've got the infantry commander on board with me—he commands about six hundred to seven hundred infantry men. I'm commanding that element of the helicopters. Well, we are trying to prep the landing zone [LZ] and get it as soft as possible so that we know we can land and aren't going to get shot up. The first thing that happens is that the Air Force comes in and strafes the area and bombs it, and we were directing them. This is such a large operation that in addition to this poor infantry commander sitting there, we've also got our boss above us in a helicopter, and his boss is above him, and above him is his boss. This was the kind of over-supervision that was so prevalent in Vietnam. As we are getting the area softened, they are giving us comments like, "You ought to put the Air Force around a little bit this way," or "Make sure you get that wood line over there." Both the infantry commander and I will be getting those kind of comments.

After the Air Force has hit the area you figure that nothing living would be there. The last thing that happens is I bring in the helicopter gun ships and they cut right down on the tree tops and they strafe the tree lines next to where you are going to land. And then they give you a report back, such as, "Hey, the LZ looks cold," which means they didn't draw any fire. Well, that isn't the report they made. After making their last pass, they reported that the LZ is hot and they took heavy fire.

It's crunch time. This infantry commander and myself need to make a decision. We've got an armada on its way in and we are about thirty seconds out from landing. It's going to be very difficult to turn around. Are we going to land or not? I look at him and he looks at me. We'd both love to hear a comment now from our bosses—"Hey, you oughta go in," or "No." But there's total silence. They don't want any part of making a decision that might go bad. Up to now they'd been second-guessing everything we'd done. "Okay, second guess us again, guys: tell us if we should land or not." Silence. Nothing.

Ron Fernstedt, Marine infantryman, Vietnam

After Operation Colorado, they evaced me, took a little bit of shrapnel out of my arm, and found out that I had malaria. I was on the hospital ship *U.S.S. Repose*. While I was on the *Repose*, we sailed around the South China Sea and stopped in Manila and Hong Kong. Then we went back to Chu Lai. I was feeling pretty good. I was only down with malaria for about a week, and I didn't have any problems moving around.

They told me I had to stay on the ship another month, so I went AWOL. I got on a landing craft, went ashore at Chu Lai, went out to Highway One, stuck my thumb out, and hitchhiked up to Hill 54. When I thought about it later, I was totally scared, but at the time it seemed like the thing to do.

I walked into the battalion headquarters and was

told, "Fernstedt. Glad you're here. You're now a company commander." They sent me out to an island in the Chu Lai Harbor. It was a beautiful place; on one end was a cliff, and the rest of the island was sand. It had beautiful white sand beaches. Below the missile site and just back from the beach, there was a village in a big grove of palm trees. Down on another corner of the island and right across from the naval base was a village. The other corner of the triangle was Hill 12; that's where my unit was. It was a composite unit of combined-action squads, security squads, machine gun squads; 106 squads. Just a bunch of people dumped there to fulfill a mission: to provide security and to run the combined-action program. It wasn't a formally-designated company, but I had mortars, medics, and everything else. I had 250 Vietnamese Popular Forces and about 160 American Marines. Every night we sent people out to patrol the channels and the other islands, to set up ambushes, and to provide security for the missile site and the Chu Lai base area.

I was able to work with the Vietnamese people through the combined-action squads, which were half Marine and half Vietnamese. Because I was responsible for the security of the area, I worked with the elders in the village. I'd get up about ten in the morning, wander out of my bunker, run on the beach for about a mile just to keep in shape, and then come back and swim out to a sand spit. By that time, Mi, my house mouse, would have a palm mat stretched out on the beach and have some cold sodas for me to drink. I'd lie there on the beach and get a suntan for the next four or five hours. During the day, we controlled the world. About four in the afternoon, I'd go in and prepare the overlays and plans for the patrols that I was sending out, brief the patrol leaders, make sure that I had all the supplies they needed, and then I'd go to sleep. I'd sleep until about seven, when the patrols would go out. From then until about three or four in the morning I had to be awake and near the radio to make sure that all the mortar concentrations were plotted and ready to fire. At about three or four, when the last patrol came back in, I'd go to sleep.

Yvonne was Vietnamese, French, and Chinese. She was educated in France and spoke several languages fluently—English, French, Chinese, and Vietnamese. She helped me with my French and Vietnamese, and I helped her practice English. She spoke English quite well. She must have been about sixty years old. She ran most of the whorehouses in the Chu Lai area. She must have had about 100 to 150 girls working for her. She had a beautiful house on the island. It was concrete-faced, with a big courtyard and a hardwood interior. She had a number of servants that cooked and did all the work. Yvonne and I became good friends. We'd sit, talk, and play chess.

We were doing all right there on the island, but then Yvonne's grandchild became sick; dying, according to my corpsman. I arranged for the child to be medevaced to the hospital at Chu Lai, and the kid recovered. After that Yvonne and I became really good friends. From that time on, the VC couldn't move anywhere near the island because Yvonne knew everything, and she relayed the information to us. She had her fingers in every pie in the area. With Yvonne's people, we had an intelligence net that was as good as anything the VC had.

The island changed drastically. When I first went there with my men, the people were scared. They had to be in their houses and quiet as soon as the sun went down. The kids were kids, but they were scared all the time. We went there and the school reopened. We taught at the school, and we worked with the people. One of my men was a rice farmer from Louisiana. He went out and helped the rice farmers through an interpreter. His dad sent him a couple of fifty-pound bags of rice seed. We became good friends with the Vietnamese people on the island, and they got to be pretty good friends with us. We helped dig wells and did some work with sanitation. We also helped with the farming. As far as I was concerned, we won the war in that little area.

Jerry L. Jensen, Army Special Forces, Korea and Vietnam

I'm sure that at first the other men had questions until we were in combat together for a while, and then they found out that even though I was religious I was still a soldier. It was only about a week after we were together as a team that we'd go out on a combat patrol and they'd ask me to have prayer for them before we left. We'd have prayer, not that we could go out and kill all kinds of people, just that we could do our duty and be able to do our job the way we were trained to do.

INDEX

A

Adams, George L., 39, 140, 155
Andrews, Donald George, 122, 233
Ash, C. Grant, 4, 106, 124, 125, 185, 194, 199, 205, 207

B

Baldwin, Richard A., 94, 99, 105
Beard, Richard Paul, 8
Bell, Peter, 5, 119
Bigelow, LaVell Meldrum, 101
Bowers-Irons, Timothy Hoyt, 26, 51, 147, 153, 171, 211
Brown, G. Easton, 187, 201
Butler, Jay Dell, 34
Byington, Freeman J., 212, 225

C

Campbell, Eugene E., 50, 213, 217
Cary, Robert G., 119, 157
Christensen, Dallis A., 56, 58, 87
Christy, Howard A., 11, 59, 163, 217, 227

D

Detweiler, Robert, 117, 141, 156, 224
Duff, John A., 89, 102, 131, 170

E

Elton, Calvin William, Jr., 49, 183, 190, 197, 201, 209
Evans, David L., 24, 27, 30, 40, 42, 127, 143, 154, 163

F

Farnworth, Ivan A., 13, 28
Farnworth, Kim, 119, 141
Fernstedt, Ron, 11, 126, 129, 167, 234
Folkman, David I., Jr., 88, 91
Foote, Danny L., 18, 21, 43, 58, 125, 126, 132, 143, 158, 163

G

Gardner, David L., 139, 167
Gunnell, E. Leroy, 10, 157

H

Haines, Albert E., 28, 133, 135, 172
Hall, Douglas T., 54
Hickman, Martin B., 16, 25, 144
Holden, Dennis, 38, 98, 130, 131
Hughes, Robert R., 6

J

Jensen, Jay R., 179

Jensen, Jerry L., 45, 57, 117, 122, 123, 125, 225, 237
Johnson, Lawrence H., 4, 9, 69, 92, 226
Johnson, Michael R., 47, 122, 146, 159

K

Kallunki, J. Tom, 127, 140

L

Le, Liem Quang, 223
Lyon, David R., 21, 145, 148, 155

M

Matheny, Ray T., 71, 74, 104, 107, 149, 191, 195, 198, 199, 202, 206
Melville, J. Keith, 3, 72, 91
Morse, George E., 118, 132, 147, 160, 164, 224

N

Norton, John, Jr., 64

P

Packer, Lynn, 7, 128, 162
Palmer, Spencer J., 30, 173
Parkinson, Edmond S., 8, 53

R

Riley, W. Howard, 173

S

Sammis, Norman Wade, 25
Speidel, Walter H., 44, 124, 181, 184, 204, 210

T

Taylor, Hyde L., 21, 53, 131, 134, 148, 166, 215
Terry, Michael B., 123, 135

V

Velasquez, Cresencio (Chris), 6, 36, 142, 160, 162, 166

W

Waldron, Kirk T., 10, 121, 130, 141, 161
Warr, Wayne A., 23, 141, 142, 158, 159, 232
Warren, Grant L., 97, 113, 129, 144, 154, 168
Watkins, Pat, 22, 29, 221
Weaver, Ted L., 14, 18, 70, 84, 99, 130
Weeks, Clyde Everett, Jr., 121, 169
Weeks, Werner Glen, 95, 222
Whitaker, Lincoln R., 32, 37, 41, 42, 46, 133, 134, 154, 156, 164, 214
Workman, Cornelius (Neil), 5, 120, 129, 165, 170, 216